Superstorm Sandy

Nature, Society, and Culture

Scott Frickel, Series Editor

A sophisticated and wide-ranging sociological literature analyzing nature-society-culture interactions has blossomed in recent decades. This book series provides a platform for showcasing the best of that scholarship: carefully crafted empirical studies of socio-environmental change and the effects such change has on ecosystems, social institutions, historical processes, and cultural practices.

The series aims for topical and theoretical breadth. Anchored in sociological analyses of the environment, Nature, Society, and Culture is home to studies employing a range of disciplinary and interdisciplinary perspectives and investigating the pressing socio-environmental questions of our time—from environmental inequality and risk, to the science and politics of climate change and serial disaster, to the environmental causes and consequences of urbanization and war-making, and beyond.

Superstorm Sandy

The Inevitable Destruction and Reconstruction of the Jersey Shore

DIANE C. BATES

RUTGERS UNIVERSITY PRESS

NEW BRUNSWICK, NEW JERSEY, AND LONDON

Library of Congress Cataloging-in-Publication Data

Bates, Diane C.

Superstorm Sandy : the inevitable destruction and reconstruction of the Jersey Shore / Diane C. Bates.

pages cm. — (Nature, society, and culture)

Includes bibliographical references and index.

ISBN 978–0–8135–7340–3 (hardcover : alk. paper) — ISBN 978–0–8135–7339–7 (pbk. : alk. paper) — ISBN 978–0–8135–7341–0 (e-book) — ISBN 978–0–8135–7342–7 (e-book web pdf)

1. Hurricane Sandy, 2012. 2. Coastal zone management—New Jersey. 3. Hazardous geographic environments—New Jersey. 4. Shore protection—New Jersey. 5. Human ecology—New Jersey. 6. Social ecology—New Jersey. 7. Environmental sociology—New Jersey. I. Title.

HT393.N5B37 2016

304.2'509749—dc23 2015012450

A British Cataloging-in-Publication record for this book is available from the British Library.

Visit our website: http://rutgerspress.rutgers.edu

Manufactured in the United States of America

For Laura, Nicholas, and all of the children of the Jersey Shore

CONTENTS

PREFACE AND ACKNOWLEDGMENTS

In a social milieu of constant media access, disasters like Sandy have provided an endless stream of images and narratives that can inform but may also be used for less noble purposes. Movies and television shows have long dramatized human tragedies for entertainment; now real-world disasters offer the same sort of vicarious thrill, albeit with real humans, real losses, and real suffering attached. I have deliberately tried to avoid such "disaster porn" in favor of largely structural explanations. At times, however, structural explanations have been complemented with the stories of individuals, such as the oral histories in the prologue, collected through the Hurricane Sandy Oral History Project, led by my colleague Dr. Matthew Bender. The personal stories included here were chosen because they personalize rather than sensationalize the Shore. At the same time, with the exception of public officials, I made the decision to obfuscate the identities of the individuals to avoid the potential for drawing unwanted attention to them.

In addition to Dr. Bender and his students, and all of the Shore residents who participated in the Hurricane Sandy Oral History Project, I would like to acknowledge the support for this project from many of my colleagues and students at the College of New Jersey. Among them are: Rachel Adler, Elizabeth Borland, Winnifred Braun-Glaude, Tim Clydesdale, Pat Donohue, Karen Dubrule, Chris Fisher, Lynn Gazley, Mohamoud Ismail, Jean Kirnan, Margaret Leigey, Rebecca Li, Michael Nordquist, Brian Potter, Ben Rifkin, Miriam Shakow, Jon Stauff, and Bruce Stout (who also served as the pilot for the aerial photographs I took that are included in this book). Lynn Gazley, Michael Nordquist, and Martin Bierbaum all read early chapter drafts and provided useful suggestions, and Margaret H. Bates helped edit the first complete draft. Discussion with the students in my environmental sociology class in spring 2013 heavily influenced how this project developed; among these, Pete Peliotis, Joanna Peluso, and Jigna Rao made important contributions to the way I approached this material. My former professors at Rutgers University exposed me to most of the theory on which this book is based, and for this I thank Patricia Roos, Lee Clarke, Karen O'Neill, Ira Cohen, Eviatar Zerubavel, and the world's best academic mentor, Thomas K. Rudel.

This particular book grew out of an item I wrote in the American Sociological Association's Environment and Technology newsletter, which was strongly encouraged by then-editor Michael Agliardo. Peter Mickulas, my editor at Rutgers University Press, read that short piece and believed in a book-length project long before I did. He deserves my warmest thanks and gratitude. Two anonymous reviewers provided very valuable feedback. Willa Speiser had the Herculean task of copy editing and improving the clarity of the manuscript; she made vast improvements in the manuscript, but cannot be faulted for problems that remain—these are mine alone.

I also extend gratitude to the Hive Mind, who helped brainstorm titles for this book: Maggie Benoit, Joanna De Leon, Kelly Dowd, Dave Harker, Lauren Heberle, Marla Jaksch, John Lang, Hannah McKenley, Jo McMullen-Boyer, Emily Meixner, Lynne Moulton, Kevin Muoio, Brian Musikoff, Juan Felipe Rincón, Paulina Ruf, Rebeca Sharpe, Christopher Sieving, Jon Stauff, Felicia Steele, Debi Stephens, Kirsten Streiff, Susan Strickland, Sara Tomczuk, Amy Tridgell, and Mort Winston.

My extended family have provided the support needed to undertake a project of this magnitude, and have shared with me much of the context and decades of experiences that I needed to be able to make sense of my adopted state; special thanks to Marge Bates, R. C. Bates, William Bates, Carol Bates, Juan Felipe Rincón, Debbie Bogstahl, Alexandra Press Maguire, Michelle Press, Joyce Malanga, Ralph Malanga II, Cheryl Malanga, Jenny Howell, and Allison Zicchinelli. Finally, I am particularly grateful to my husband, Ralph, and my kids, Laura and Nicholas, who have served as cheerful companions and research assistants, sacrificing weekend after weekend, as well as much of their summer vacations since Sandy, for "fieldwork" on the boardwalks, beaches, and neighborhoods of the Jersey Shore.

Superstorm Sandy

Prologue

Down the Shore, (Not) Everything's All Right

In October 2012, much of the Eastern Seaboard was preparing for a late-season storm that had intensified to become a hurricane while lingering south of Jamaica on Wednesday, October 24. Along the Jersey Shore, residents prepared as they always have: filling up gas tanks, purchasing batteries and food staples until stores were bare, storing patio furniture and flowerpots. Waterfront homeowners secured windows and boats; those who were second-home owners locked up securely and headed home to ride out the storm elsewhere. Year-round residents prepared emergency kits containing all of their personal and legal information, just in case they had to be evacuated from vulnerable areas. Shore residents had practice: Hurricane Irene had made landfall just over a year before, and many had evacuated in its path. Atlantic City even closed its casinos for just the third time since they had opened in 1978. Irene lashed the region with high winds and caused river flooding, but most of its major damage was elsewhere, as the storm released record-setting rains in New England. There was little reason to think that Sandy would be any different.

As the storm grew nearer, Ocean County residents began to take the storm a little more seriously. Hurricane Sandy had grown and intensified, and most weather models predicted landfall somewhere on the Jersey Shore. In an interview one year after the storm, Mayor Thomas Kelaher of Toms River recalls the warning given by a representative of the National Weather Service: "Mark my words, there is nobody alive in the State of New Jersey who has ever seen what we are going to see when this storm comes ashore."[1]

The region's emergency management officials and first responders increased pressure on residents to evacuate, while the residents themselves scrambled with last-minute preparations. A stubborn few refused the evacuation order and continued to move furniture and valuables to higher locations within their homes. They brought blankets, flashlights, candles, nonperishable food, and pets into their attics, or decided to relocate to more elevated homes of friends and neighbors. Everyone paid close attention to their televisions, computers, and phones, which streamed nearly constant coverage of the impending storm.

By Sunday, October 28, most of the residents in mandatory evacuation areas were gone and only officials remained.[2] Louis Amaruso, the director of public works for Toms River, crossed over the Barnegat Bay to check on the township's beachfront neighborhoods. Although landfall was still nearly thirty-six hours away and the storm was hundreds of miles off shore, Amaruso remembers, "I went up to see how the secondary dune line was doing and it was already gone. It had already been taken out. The permanent dune line was starting to be eroded and [one house] had already lost its deck and water began rushing through the house. The house was starting to come apart." He told his companion, "Take a good look around because none of this is going to be here, it's all going to be gone. This whole area will be lucky if there's an island left after this."[3]

Up the coast in Mantoloking, the police force was still patrolling this nearly deserted town of large, expensive second homes. Officer John David Barkus spent a long night helping prepare for the storm; by 5:30 the next morning (Monday, October 29, fourteen hours before landfall), he reported the first breach of the dune line in Mantoloking, and water flowing freely across State Highway 35 shortly thereafter. By 8 AM, he observed that the protective dune was completely gone and waves were crashing onto the decks of houses. An hour later, Barkus met with a caretaker of a seasonal home and recounts: "We were standing in the living room and the waves are crashing on the bay window on the ocean side of the house. So at that point there, we know we were in for a big event." At around 3 PM, Barkus received a call from his sergeant, who had been "hit by a wave on Route 35" and was told that even emergency personnel would be evacuated across the Mantoloking Bridge to the mainland in Brick Township.

Sandy made landfall north of Atlantic City at 7:30 PM, while the storm's northeastern quadrant continued to batter the Ocean County shoreline. The Point Pleasant Beach Weather Center recorded a gust at eighty-four miles per hour before its instrument was ripped away by the wind. Sometime between 10 and 11 PM, the ocean punched through the barrier peninsula in Mantoloking and water rushed into the Barnegat Bay, which rapidly rose six feet or more, flooding up inlets, lagoons, lakes, and rivers on the mainland, well beyond the evacuation zones. Toms River Police Captain Steven Henry later noted water rescues "two or three miles inland." Councilman William Robert Mayer explained that in Point Pleasant Beach, "west of the tracks, most of the damage was from the [Manasquan] River, east of the tracks it was a combination of the river, the lakes, and the ocean." Harry "Chip" DeCorsie, a Point Pleasant Beach Office of Emergency Management (OEM) dispatcher, described the office's phones as "ringing off the hook" with people needing assistance. Toms River's OEM coordinator,

Paul Daley, estimated 3,200 emergency calls: "We had people calling that had kids in their attic. Their houses were filling up with gas and we couldn't get to them." Bob Burger of the Point Pleasant Beach Weather Center offered the following advice: "Bunker down, hunker down, and wait for the lights to go out. Hope that a tree doesn't fall on your house."

Despite high levels of compliance in the evacuation zones, a record high tide overnight thwarted rescue efforts even on the mainland. In Point Pleasant Beach, Deputy Emergency Management Coordinator in Charge of Operations Kyle Grace explained that even high-water rescue vehicles were unable to reach people calling for help. An attempt to cross the Toms River Bridge resulted in an army vehicle nearly floating off into the Barnegat Bay. Toms River Public Works sent front-end loaders into the flood waters at the height of the storm, despite howling winds and no electricity, eventually completing more than five hundred rescues on the mainland overnight. For people remaining on the barrier peninsula, first responders could do little more than record the addresses of callers, with a promise to get to them as soon as possible. Overnight, they huddled in their upper floors and attics, seeing the ocean wash away their neighbors' houses, hearing the wind tear off roofs, and watching the water rush into their homes from under the door, then through heating vents and electrical outlets.

After helping rescue twenty-two people from flooded vehicles in Brick on the mainland, Mantoloking police officer John Barkus tried to catch some sleep on an air mattress in the attic of a local business. Unable to rest, at 4:30 AM he joined a group of Brick Township police officers and volunteer firefighters who attempted to cross over the Mantoloking Bridge. He describes what he saw: "I could see waves obstructing, coming from the ocean on the north side of the bridge going from the ocean right into the bay—five- or six-foot waves going through. There was a wall of debris on the east side of the bridge seven to eight feet tall. . . . We watched [a] house come off its foundation, flow out into the water, and crash into the bridge while we were on it." They retreated from the bridge but eventually convinced two Ocean County Public Works employees to drive them to Point Pleasant Beach on a front-end loader to assist in recovery efforts there.

As daylight returned to the Jersey Shore on Tuesday, October 30, the damage was worse than people had imagined. First responders and Good Samaritans entered flooded areas on watercraft and helicopters, carrying stranded residents to safety before heading back to find more. Water receded slowly, and residents who tried to return to their flooded homes spoke of wading through chest-deep water, filled with debris, fuel oil, and "Lord knows what else." Fish swam in the water, or lay dying or dead on dry land; one resident reported seeing harmless sand sharks swimming in the streets of Seaside Heights. Lawns far inland were covered in flotsam—not just small debris, but chunks of buildings, docks, and entire boats. A hot tub sat in the Point Pleasant Beach train station parking lot. Foul, menacing smells of muck and natural gas persisted, despite strong winds from the retreating storm. Residents reentered their homes to find massive water damage, as well as sand and mud throughout lower floors. Floodwaters had overturned refrigerators and freezers, bins of dog food, and fuel oil tanks, contributing to the stench of disaster.

Near the ocean beach, "nuisance does not adequately describe the magnitude of the sand," said Point Pleasant Beach councilman William Robert Mayer; the sand was described by many as being like snow after a blizzard. In fact, sand was removed from major highways with snow plows and consolidated into massive dunes. Firefighter Robert S. McIntyre, who also serves as Mantoloking's OEM coordinator, describes the return to his home on the barrier peninsula: "There was six feet of sand in our driveway—like [the driveway] had never been there—but you could hear the gas shooting up from the mains. So first of all, I started with the gas company. I told them to shut off Mantoloking and I said immediately. They said, 'We can't do that . . . because if we depressurize the system, there will salt water intrusion and we'll lose our system.' And I said to them, 'You don't have a system, so turn it off.'" Such advice came too late for the community of more than one hundred homes in Camp Osborne on the Barnegat Peninsula in Brick Township, which all caught fire and burned to the ground, casting an ominous glow that was noted by first responders miles away in Point Pleasant Beach.

Mayor Thomas Kelaher describes his first trip across the Toms River Bridge to the Barnegat Peninsula in detail:

> We got out and walked up by where the boardwalk was gone, and I mean, literally. . . . Down one of the side streets from the ocean, we saw a chunk of blacktop up against a wrecked building, and it had part of the blue handicap decal which had been up by the boardwalk. Three years ago, we had dedicated the opening of a brand-new lifeguard tower at the south end of Ortley Beach, public bathrooms opening up onto the boardwalk, three-story-deep pilings, two-foot concrete slabs, masonry walls, steel stairways, and we went to see how that held up. It wasn't damaged; it was gone, just plain gone. And down the street later we found chucks of the masonry, which we could later identify because it had a yellowish hue to it, and that was it. Police told me weeks later they found the roof of the lifeguard tower on top of the Chinese restaurant down on the highway.

More troubling to the mayor—and residents who trickled back into the area, often without official sanction—were the number of homes that were destroyed or had simply vanished. "The people who left for a couple days thinking it would be like Irene, only left with the shirts on their backs," worried Mayor Kelaher. "These people aren't going to be able to go back, there's nothing to go back to." Residents and officials cite various engineering reports that indicate that from Mantoloking to Toms River, waves higher than thirty feet had scoured the coast. Houses that were carried by these waves smashed into houses farther inland, with sand erasing all evidence of their original footings. When the water receded, some of homes were left in awkward positions, listing against other structures or abandoned in the middle of Route 35. Other homes and all of their contents disappeared entirely into the ocean or the bay, although detritus from the Barnegat Peninsula was later recovered miles from its point of origin, up and down the beach as well as on the west side of Barnegat Bay.

Trying to capture the damage in words, residents turned to comparisons far from their everyday experiences. Residents compared the landscape to movies, Ground Zero, or a "war zone" where someone had set off a bomb. Officials trained in emergency management echoed these sentiments: the devastation was "literally like a movie," "like Beirut or Iraq or Afghanistan"; they made comparisons to Louisiana after Katrina. Captain Steve Henry of the Toms River Police Department asked, "Have you ever seen the movie *Planet of the Apes*? Not the remake, the *real* one. The actor walks onto the beach and [he sees] the Statue of Liberty [buried in sand]. That's exactly what we thought it was; it was like Armageddon and the end of the world. It was insane, it really was."

And yet, reconstruction began almost immediately, even in those communities that had been most devastated by the storm. The Jersey Shore quickly proclaimed itself on T-shirts, bumper stickers, and billboards to be "Stronger than the Storm" and vowed to "Restore the Shore." Tens of thousands of volunteers turned out to demonstrate commitment to these sentiments and help rebuild. The federal, state, and local governments—as well as countless nongovernmental organizations—pledged billions of dollars of recovery assistance.

Reconstruction is designed—more or less—to take into consideration that storms like Sandy are likely to hit the Jersey Shore again. After all, Sandy was not the first storm to do major damage to the Jersey Shore. Shore residents, most of whom have a healthy respect for the power of the ocean based on firsthand experience, understand that "when nature wants to kick your ass, it's going to kick your ass, whether you're ready or not." Why, then, did we collectively put these communities at risk in the first place, and how can one explain this Sisyphean commitment to rebuild?

This book attempts to provide insight into this query, drawing upon both the unique character of the Jersey Shore and more universal aspects of the way that humans structure their relationship with their environment. Unlike other living things, people don't just occupy habitats, they inhabit places rich with social, cultural, and personal meanings. People assign both emotional and financial value to places; political and economic systems reflect and reinforce these values. Some social groups are better positioned to assert their interests, sometimes at the expense of other social groups and the nonhumans who share our habitat. These patterned social processes have influenced how people altered New Jersey's coastal habitat before October 2012 and have influenced the process of rebuilding since. Sometimes, these are based in the physical or historical specificity of this place; other times, they reflect broader patterns of social organization that have been observed and explained elsewhere. As such, the goal of *Superstorm Sandy* is to provide insight into the fundamentally precarious environmental position we humans have created—both here in New Jersey and across the planet.

1

The Inevitable Sandy

New Jersey is a much-disparaged state, especially for its environment, which has been subject to the excesses of its human inhabitants since the colonial era. Despite its reputation as an industrial wasteland, New Jersey is an excellent place to study the interactions between modern humans and our natural environment in an array of contexts beyond contamination. Although none of its diverse ecosystems are free of the influence of human populations, New Jersey has a long history with agriculture (hence its nickname the "Garden State"), forest preservation and conservation (for example, the Pinelands Reserve), the protection and restoration of wetlands (for example, the Hackensack Meadowlands), the preservation of surface water quality (for example, the Highlands Preservation Area), and coastal land management (for example, the Coastal Area Facility Review Act). New Jersey is also an excellent location in which to study the effect of natural disasters, as its coastline is regularly subject to both winter storms (nor'easters) and tropical cyclones (hurricanes and tropical storms).

This is the story of one of these storms, Sandy, which made landfall at 7:30 PM in Brigantine on October 29, 2012, traveling north up the coast from Atlantic City, shortly after being downgraded from an official hurricane. Alternately referred to as Hurricane Sandy, Superstorm Sandy, the post-tropical-cyclone-formerly-known-as-Sandy, and (according to the state's largest newspaper) Frankenstorm, Sandy was the most severe coastal storm to hit the Jersey Shore since the Ash Wednesday Storm of 1962. A total of 147 people in the Caribbean and United States (12 in New Jersey) died as a direct result of the storm, which has now been estimated to be the second costliest hurricane in U.S. history after Katrina.[1]

In comparison to Hurricane Katrina, which devastated New Orleans and the Gulf Coast in 2005, Sandy was not particularly strong. The National Weather Service had downgraded the storm to a post- or extra-tropical cyclone two and half hours before its eye crossed the Jersey coastline,[2] although later analysis of meteorological records suggest that sustained hurricane-force winds of at least seventy-five miles per hour did reach large swaths of the Garden State. Rainfall was much lower than even in recent storms—just the year before, Hurricane Irene had flooded the state with up to twelve inches of rain, while Sandy dropped only between five and twelve inches, with more rain falling *south* of the storm's eye and away from the most devastated parts of New Jersey; the highest rainfall total of 11.91 inches was recorded in Wildwood Crest.[3]

Nonetheless, Sandy was a spatially immense system with forty-mile-per-hour sustained winds ("tropical storm winds") extending for more than nine hundred miles.[4] These sustained winds pushed surface waters in the Atlantic Ocean into a storm surge to the eye's north and east; coincidentally but unfortunately, Sandy's landfall corresponded with a lunar (astronomical) high tide, which generated an unprecedented storm surge up the coast of New Jersey and into New York and Connecticut. The highest official offshore surge, 8.57 feet, was recorded off Sandy Hook,[5] but since the weather station buoy stopped functioning shortly after making this measurement, the National Hurricane Center assumes the actual surge was higher. As water squeezed into the region's bays and inlets, it rose. At the U.S. Coast Guard Station on Sandy Hook, the surge rose to 8.9 feet above *ground* level. The wall of water spread along the Raritan Bay shore of New Jersey, where high-water marks were indicated at 7.9 feet in Keyport and 7.7 feet near the mouth of the Raritan River in Sayreville. Water funneled north into New York Bay and backed up the Hudson and East Rivers, flooding trains, tunnels, and waterfronts. High-water marks ranged from 6 feet in Manhattan's Battery Park, to 6.5 feet in Hoboken, New Jersey, to eight feet at the Brooklyn Bridge. The south shore of New York City's Staten Island undoubtedly suffered the most in terms of human losses: twenty-one people perished and entire neighborhoods were eliminated overnight.

Acknowledging the human deaths caused by the storm, damage in New Jersey was mainly limited to property damage. Early estimates indicated that more than 346,000 homes were damaged or destroyed, representing nearly one in every ten residences in the state.[6] Nearly 19,000 businesses reported at least a quarter-million dollars of damage, and total economic losses were estimated to be over eight billion dollars.[7] Power outages were widespread, affecting five million residents, many of whom lost power for weeks afterward (my own power was out for "only" six days). Unsurprisingly, the most heavily damaged communities were those along the water, especially on the barrier islands and along the Bayshore, which lies across the Raritan Bay from the south shore of Staten Island. Advance warning, well-organized evacuations, and borderline bullying from Governor Chris Christie had been successful in getting most of the residents of these areas out of harm's way. The area also had much lower populations to evacuate than would have been the case if the storm had hit during the summer season. Even so, the damage to the Shore was unimaginable.

Three iconic images emerged to encapsulate the damage done by the storm, and they have motivated restoration efforts on the Jersey Shore. The most widely publicized was the Star Jet rollercoaster, which had fallen into the Atlantic Ocean off Seaside Heights, when the Casino Pier collapsed beneath it. The second portrayed the Mantoloking Bridge, which had once connected Brick Township to the Barnegat Peninsula, and after the storm, ended at a new inlet, surrounded by piles of rubble—the remnants of multimillion-dollar homes. Finally, images of more modest homes swept away at the foundation portrayed the devastation of the Bayshore communities, particularly in Union Beach. Each image captured a different type of property damage (commercial enterprise, infrastructure, and personal property of both the wealthy and the not-so-wealthy), which, along with the widespread power outages and loss of trees, made the storm feel very democratic in its impact.

People in New Jersey—and across the United States—responded quickly with emergency aid, stemming from both the governmental and the nongovernmental sectors. Although there were some hiccups, in comparison to Katrina emergency response to Sandy along the Jersey Shore was very well coordinated and executed. The governor and President Barack Obama had declared the region a natural disaster area before the storm had even made landfall, and

1.1. The Star Jet roller coaster offshore from Seaside Heights, winter 2012. Photo by author.

1.2. Mantoloking Bridge from above, fall 2013. Photo by author.

both were on the ground touring devastated communities within forty-eight hours. The National Guard was called up both to provide direct relief effort and to protect depopulated areas from looting. Municipalities and counties worked in concert to provide immediate assistance, facilitated by electronic and social media, which made it possible for shelters to address even very specific needs. For example, two days after the storm, I read a Facebook post that the shelter at Toms River North High School no longer needed food or water, but still needed batteries, socks, and diapers. As damage assessment continued, government actors at the local, state, and federal levels continued to have relatively good relations with affected populations, especially in comparison with the debacle following Katrina.

As time has passed, affected residents have grown increasingly disillusioned by government, particularly at the federal level. Republican governor Christie had fairly consistent and impressive support among the predominantly independent and Democratic Garden State residents (at least until the "Bridgegate" scandal broke in early 2014). His approval rating was enhanced when he broke ranks and criticized the Republican national leadership for delaying emergency aid in January 2013. Christie was a constant presence in the media spectacles associated with the storm's aftermath, hanging out backstage with Bruce Springsteen, Jon Bon Jovi, and Steven van Zandt at the 12.12.12 benefit concert, presenting a signature Governor Christie fleece jacket to England's Prince Harry before he toured Mantoloking and Seaside Heights in May 2013, and even winning a stuffed teddy bear for President Barack Obama during his visit to Asbury Park's boardwalk later that month. In contrast, dissatisfaction with the federal government and its representatives has grown, including ire over the blocked federal aid in January 2013, frustration with flood maps and elevation requirements, and the perceived inadequate and capricious financial support from FEMA and other public agencies.

Apart from some environmental organizations, most vocally the New Jersey chapter of the Sierra Club, and in contrast with New York, very few people have questioned whether rebuilding should occur in the affected areas. Instead, the state draped itself in "Restore the Shore" paraphernalia—hats, shirts, bumper stickers, magnets, icons—and actively set about doing just that. Armies of volunteers from throughout the state (and, it should be noted, from all over the country) have devoted thousands and thousands of hours to cleanup, demolition, and rebuilding. Contractors and engineers have been busy; the rebuilding of Sandy's destruction may pull the state's construction industry out of its five-year slump. The rush and euphoria of rebuilding the Shore have waylaid a deeper discussion of why Sandy was such an expensive and destructive storm, thus avoiding careful consideration of whether rebuilding the Shore will be a Sisyphean task, with all that restoration to be wiped out again by future storms, which many climate models predict will grow stronger and more frequent.

Indeed, restoration has been approached mainly as a technical problem: how to rebuild homes, roads, electrical lines, and bridges in such a way as to prevent—or at least minimize—damage from inevitable future storms. Despite mounting frustrations and increasing resignation, few have stopped to consider why damages occur; why, how, and where they did; why technological solutions remain controversial; why government can't just "fix" things; why some people receive aid and others do not; and a host of other unresolved questions associated with the storm.

This book offers insight into these questions through key concepts from environmental sociology, a social scientific perspective that seeks to identify and explain patterns in modern humans' interactions with their environment. Environmental sociology explains why, despite unprecedented preparation in advance of the storm, Sandy caused so much destruction and why restoration efforts have taken the form that they have. After all, if Sandy had struck an undeveloped coastline without resident human populations, the story of this storm would be very different. Sociology and other social scientific disciplines are often overlooked when considering environmental problems, even though the root causes of most environmental problems are not "natural" or technical, but social.

Environmental Sociology as a Lens to Understand the Jersey Shore

C. Wright Mills called sociology the "intersection between history and biography." By this he meant that sociologists study how individuals and the circumstances in which they live influence one another. According to sociologists, individuals are shaped by social structures that are beyond their individual control but are shared by other members of their society. Individuals do have the ability to carve out unique histories, trajectories, interpretations, and experiences ("agency"), but they do so within the institutions and cultural patterns that are available to them in their time and place ("structure"). Sociologists use the scientific method—the systematic collection of empirical data to develop and test ideas about how one thing affects another—to study this intersection between agency and structure.

Environmental sociology posits a two-way relationship between individuals and their social and environmental circumstances, whereby what individuals think and feel, and how they act depends on agency (for example, personality and internal motivations), structure (for example, families, religions, political and economic systems, and technologies), and nonhuman context (for example, ecology, geography, climate). Environmental sociologists now have theoretical insights based on data collected over five decades, although no single theoretical tradition (or "paradigm") has emerged as sufficient to explain all features of the human-environment relationship. This is not surprising given that the broader discipline of sociology has wrestled for more than two centuries with three central paradigms.

In short, sociologists, including environmental sociologists, typically describe the way that society works in one of three ways. Structural functionalists believe that societies are held together by common values and interdependent institutions, such as economics, politics, religion, family, and education. This perspective derives from the classical theoretical works of Emile Durkheim. Conflict theorists, in contrast, propose that societies are organized around hierarchies. Groups at the top control political, economic, and value systems for their own benefit, while other groups must fight for every concession. This perspective draws from the classical theoretical tradition associated with Karl Marx and Friedrich Engels. Finally, constructionists believe that society is best understood through the meanings or understandings that people develop during socialization and interaction. These meanings can become so customary that people forget their origins and behave as if there were no other alternatives (that is, they are "naturalized"). Constructionism has its origins in the writings of Max Weber, but also reflects contributions from early American sociologists like George Herbert Mead and Charles Cooley and contemporary European social theorists including Anthony Giddens and Pierre Bourdieu. Although most individual sociologists tend to rely on one of these paradigms more than the others, the discipline as a whole patently refuses to choose among them. Sociologists recognize that all three perspectives have withstood the tests of time and empirical evidence. Each paradigm highlights different aspects of social organization, just as visual light, infrared, and x-ray photography all are accurate records that simply highlight different aspects of reality.

Environmental sociology emerged as a subdiscipline in the early 1970s, as an academic arm of the "second wave" of U.S. environmentalism that brought the country a variety of environmental benefits, including Earth Day and the Environmental Protection Agency (EPA).[8] Early environmental sociology tended to focus on modern value systems, and specifically, how our cultural beliefs peripheralized the environment. In 1978, William R. Catton Jr. and Riley E. Dunlap published their seminal article, "Environmental Sociology: A New Paradigm," in the *American Sociologist*, which deliberately and decisively distinguished environmental sociology from the rest of the discipline. What makes environmental sociology unique, according to Catton and Dunlap, is that it rejects what they called the "Human Exceptionalist Paradigm (HEP)" in favor of a "new environmental paradigm (NEP)." A century before, Charles Darwin had written that species evolved as a response to constraints in their ecosystems. HEP asserted that humans were no longer subject to the natural law of evolution. If we were threatened by predators, we hunted them out of existence. If our environment failed to provide enough food, we developed agriculture. Humans could reduce or eliminate the impact of natural disasters with better building materials and early warning systems. As a discipline, sociologists had largely embraced the HEP, which essentially meant that humans were able to transcend natural constraints because we are a technological and innovative species. In contrast, Catton and Dunlap offered the NEP as a world view that treats humans as part of our own ecosystem, albeit unique in our ability to reflect on and therefore change the relationship that we have with it.

By the time Catton and Dunlap published their article, both academics and public intellectuals had already begun to question whether humans were really excepted from natural limits; biologist Rachel Carson's best-selling *Silent Spring* (1962) had already popularized the knowledge that chemical use (notably pesticides) had detrimental and irreversible consequences for nonhuman species. Another biologist, Paul Ehrlich, described how human population growth could easily outstrip natural resources in another best seller, *The Population Bomb* (1968). In 1969, the Cuyahoga River through downtown Cleveland caught fire and spurred many of the first federal antipollution statutes, including the Clean Water Act of 1972. Popular fiction of the time included *Ecotopia* (1975), Ernest Callenbach's novel about an ecologically sound alternative future in the Pacific Northwest, and Edward Abbey's *The Monkey Wrench Gang* (1975), which featured a band of rogue militant environmentalists bent on destroying the damaging energy infrastructure of the Southwest (and eventually helped crystallize a radical environmental movement in Earth First! in 1980). All of this is to recognize that environmental sociology emerged as part of a broader intellectual movement that doubted human exceptionalism and was diligently collecting the evidence necessary to demonstrate how humans had to reorganize their societies to live within natural limits or face a dramatically degraded environment.

In addition to Catton and Dunlap, an early cohort of environmental sociologists developed key ideas about how humans interacted with their environment. Kai Erikson, a sociologist at Yale University, researched and wrote about how a flood along Buffalo Creek in West Virginia had disrupted community and psychosocial security. Adeline Gordon Levine documented the response to toxic contamination buried beneath a residential subdivision in the Love Canal neighborhood of Niagara Falls, New York. In 1980, Catton published *Overshoot*, which explicitly linked social and political problems to ecological origins. Also in 1980, Allan Schnaiberg of Northwestern University published *The Environment: From Surplus to Scarcity*, anchoring environmental sociology in a Marxist tradition by detailing how structured inequality and growth, both of which are endemic to capitalist political economy, produce and reproduce threats to the environment.

Organization of Superstorm Sandy: The Inevitable Destruction and Reconstruction of the Jersey Shore

Over the next few decades, environmental sociologists developed a series of informative theories about the human-environment relationship, and have backed these up with considerable empirical research. Environmental sociology has not reduced itself to a single theoretical tradition, but maintains both complementary and competing ideas about how people interact with the environment. This book draws from several of the most prominent of these ideas to reexamine the story of Sandy on the Jersey Shore. Although all chapters consider insights from multiple traditions, each chapter adopts a primary theoretical lens to explain the social circumstances leading up to the storm and the social responses in the immediate and longer-term aftermath. For readers more familiar with environmental sociological theory, these ideas are not new. For most readers, however, the following section provides an overview of the main theoretical perspectives employed in later chapters.

Sociologists have long understood that "reality" is constantly interpreted by people, and how people assign value and meaning is a product of social learning and shared understanding. Chapter 2 begins to consider how people think about or "construct" their relationship with the environment. Environments are deeply imbued with social and cultural meanings; some of these are deliberate and others are subconscious. Both are relevant for examining the context of the Jersey Shore leading up to Sandy and after Sandy. The Shore is a cultural touchstone in a state that lacks a central city or even sports team to build a shared identity among the state's residents.[9] Benjamin Franklin once described New Jersey as a "keg tapped on both ends," with the north linked inextricably to New York City and the south to Philadelphia. This regionalism continues to divide the state economically and culturally, although the emergence of a "Central Jersey" identity in the past few decades increasingly confounds this division. But both North and South Jersey identify with the Shore (historically with different subregions along the Shore). This chapter thus begins with the Shore as part of the state's cultural iconography, and importance in establishing identity for New Jersey residents, including those who live far from the Shore itself. As a resort area, albeit one that primarily attracts people from within the state and from nearby New York and Pennsylvania, the cultural importance of the Shore is much more widespread in the state than outsiders may guess.

The Jersey Shore is mainly a vacation destination for families, and despite its recent notoriety, many beach towns limit the sale of alcohol or actively discourage a rowdy nightlife. Because of its importance in family vacations across generations (including second homes), New Jersey residents connect to the Shore through shared experiences of playing on the beach as children and enjoying boardwalk amusements, often with members of their extended family and usually at locations that are "traditional" to the family history. The Shore is also a place to mark important milestones; high school seniors (mostly from North Jersey) traditionally spend "senior week" at the Shore, and, throughout the state, high school proms are often followed (and sometimes preceded) by parties at the Shore. Likewise, Atlantic City's casinos and nightclubs have become a destination for New Jersey residents on their twenty-first birthdays. The cultural connection to the Jersey Shore is consistent with what sociologist Kari Norgaard calls a "double reality" in which people are deeply and emotionally connected to an environment that they know is intimately linked to insecurity.[10] Knowledge of historic storms along the Jersey Shore is widespread and easily accessible to anyone interested; at the same time, residents can deny these threats in their everyday interactions with the coastal environment because these disruptions do not jibe with the dominant construction of the Shore, which is rooted in the security of family.

To explain this, I turn to John Hannigan's ideas about constructing environmental agendas, with a particular emphasis on public narratives about the Jersey Shore that highlight the dominance of human ingenuity over nature (for example, the state's "Stronger than the Storm" advertising campaign), a therapeutic community (for example, "Jersey Strong"), and a complete failure to consider whether rebuilding should even occur (for example, "Restore the Shore"). These narrative frames are clearly rooted in emotions and iconography, including leisure, youth, summer, and family, and restoration has created an opportunity to assert these over less desirable but competing narratives, such as the *Jersey Shore* reality television show, corruption, and perceptions of overall seediness and dirtiness. In the wake of Sandy, these cultural ideas (or what Hannigan calls "moral frames") became important for justifying reconstruction without any critical consideration of the near-certainty of future storms. To deny reconstruction is to deny the cultural continuity of the Jersey Shore across generations, not a reasonable response to coastal overdevelopment. This is consistent with what Norgaard calls a "culture of denial" about likely future storm risks. It is clear that future risk has been an important part of the restoration effort, but the discussion of climate change, which will likely increase the frequency and intensity of storms like Sandy, is rarely part of the restoration narrative.

While cultural sociologists like Norgaard highlight how common understandings serve to unify social groups, other environmental sociologists point out that modern society is riven with inequalities. Chapters 3 and 4 of this book develop the environmental sociological tradition anchored in analysis of political economy and inequality. Northwestern University sociologist Allan Schnaiberg located the root cause of environmental degradation in a capitalist system that required growth to persist. Limitless economic growth, or, as Schnaiberg called it, the "treadmill of production," is entirely incompatible with sustainability in a finite ecosystem. Economic logic thus becomes the prime mover of environmental degradation, rather than a potential solution, which is the suggestion in ecological modernization theory. In reference to a nonindustrial economy like the Jersey Shore, the treadmill of production is manifest in what political sociologists David Logan and Harvey Molotch refer to as a "growth machine," in which local business owners, politicians, real estate developers, and residents develop a vested common interest in the expansion of economic activities that increase the value of their property. Environmental sociologist Thomas K. Rudel employed the growth machine model to explain deforestation in the Ecuadorian Amazon in his 1993 monograph. To Rudel, benevolent and widely shared interest in economic growth may compel the creation of environmentally unsustainable regional economies. The growth machine is the main focus of chapter 3, with an emphasis on the development and eventual decline of seaside resorts. This theoretical contribution explains both why the Jersey Shore was developed intensively prior to Sandy and why restoration is an imperative in the poststorm period—particularly in tourism destinations.

The conflict tradition in environmental sociology also highlights social inequality, whereby the poor receive the least benefit from the political and economic systems and the well-off receive disproportionately more. Environmental sociologists Robert Bullard and Paul Mohai clearly established that socioeconomic class and, persistently, race independent of class are convincing predictors of environmental quality and risk. A mountain of research amassed over decades demonstrates how siting of toxic facilities, exposure to industrial pollutants and disasters, and response to environmental threats vary by race and class, with poor and nonwhite populations bearing a disproportionate environmental burden. Bullard called the pattern whereby racial and ethnic minorities are exposed to higher levels of environmental risk "environmental racism," even though the process through which risk accumulates may be nominally colorblind. More broadly, "environmental justice" refers to both an intellectual and a social movement designed to level exposure to environmental risk, taking into account multiple potential social statuses (such as race, ethnicity, class, gender, nationality) and the intersection between them. Lisa Sun-Hee Park and David Pellow have since followed up on environmental justice to include attention to what they call "environmental privilege," or the process by which people with higher social status are able to capture resources designed to protect and enhance their environments, which produces environmental injustices at the lower end of the stratification system.

Environmental justice is used to explain how class and race are relevant on the Jersey Shore, and how environmental risks and benefits have been distributed. The suburbanization of Monmouth and Ocean Counties is an important part of understanding social inequality along the Shore. As New Jersey's urban cores declined rapidly in the wake of deindustrialization and urban unrest in the 1970s, jobs spread to edge cities (and what sociologist Robert Lang calls "edgeless cities") and made Shore communities attractive for predominantly middle-class white families. Suburbanization is linked to both white flight and the expansion of middle classes in the postwar era, but it transformed once-rural Shore counties into nearly continuous sprawl, transforming and degrading the region's environment and increasing residents' vulnerability to coastal storms like Sandy. At the same time, the Shore has become highly segregated, with poor and nonwhite communities clustered in older resorts and inland communities. In the twenty-first century, the Shore has also become been a destination for low-skilled immigrant laborers, mainly from Central and South America, who perform many supporting jobs in the tourism industry (such as kitchen help) and suburban living (such as landscaping services). The Shore region has become a highly segregated landscape, in which political and economic power has spatially and socially isolated people at the lower end of the stratification system. Although the largest and poorest minority communities of the Shore did not face the worst of Sandy's storm damage, the allocation of public funds to restore private property and recreational facilities has redirected public resources to nonpoor and mainly white residents. Sandy will likely intensify the concentration of desirable properties and environments among economic elites, as people who are unable to absorb storm losses sell their damaged homes to outsiders able to afford the costs of reconstruction.

Another major theoretical tradition in environmental sociology is introduced in Chapter 5. Ecological modernization, which is most closely associated with the substantial work of Arthur P. J. Mol and Gert Spaargaren, suggests that because people increasingly value environmental health, environmental damage is increasingly costly. In the past, harm to the environment and human health was "externalized," which meant that it was not calculated into the true costs of production; thus factory owners had no financial obligation to pay for the human health and environmental problems associated with pollution, nor was there any economic incentive to do anything other than find the cheapest way to produce goods and services, regardless of the impact on the environment. Likewise, the benefits that the nonhuman environment provides are calculated as having no economic value, but are treated as "free" services in a cost-benefit calculus; so when a forest scrubs away atmospheric carbon dioxide or filters contaminated industrial waste water, its services are not measured or assigned monetary value. As such, a cost-benefit calculation when environmental costs and benefits are externalized actually provides an economic incentive to degrade the environment.

However, as environmental science has matured and people have come to understand the value of environmental and human health, the cost-benefit calculus has shifted so that environmental damage becomes more expensive. People have come to value environmental quality for a variety of reasons. Some begin to view it as intrinsically valuable, particularly as it becomes more threatened or scarce. Others begin to hold degraders responsible for damage, both to the environment and to human health, pain, and suffering caused by environmental damages. Laws and political structures have subsequently been amended to attend to occupational health (for example, Occupational Safety and Health Act), to regulate emissions (for example, the Clean Air Act and the Clean Water Act) and responsibly dispose of waste (for example, the Resource Recovery and Conservation Act), control known toxins (for example, Toxic Substances Control Act and Federal Pesticide Control Act), protect endangered landscapes and habitats (for example, the Endangered Species Act and the Wetlands Protection Act), and determine how land can and cannot be used (for example, through local zoning ordinances). As some firms within a given industry begin to reduce their environmental impact, they develop a competitive economic advantage (as well as a moral advantage), and thus compel other firms within the same industry to adopt increasingly efficient and environmentally benign production processes and technologies. This process of "ecological modernization" is probably best demonstrated by the conversion to catalytic converters in automobiles, which has dramatically reduced some types of air pollution and improved air quality across the country (although the gains of catalytic converters has been erased by the increasing number of vehicles on the road).

It is important to note that ecological modernization does not rely on simplistic rational choice theory; ecological modernization theorists have been attentive to the fact that cultural values increasingly favor environmental protection, and thus the cost-benefit calculus is neither static nor strictly economic. Ecological modernization typically results in two types of responses to environmental damage: regulation that limits or prevents damage, and technology that minimizes environmental impact while using resources more efficiently. To date, ecological modernization theorists have mainly focused on industrial production and consumption of physical products. However, the same basic tenets apply to the Jersey Shore and its residents' approach to their environment. New Jersey residents have repeatedly enacted rules and legislation designed to protect the ocean, beaches, dunes, bays, and rivers from human impact, including legislation such as the Coastal Area Facility Review Act (CAFRA), storm water regulation, and public beach access. The economic value of clean water and clean beaches is apparent in a state where as much as a quarter of its revenue comes from Shore-oriented tourism. Storm mitigation and preparation has also been incorporated into the cost of economic activities at the Shore. Beach towns pay for dune restoration, beach stabilization and replenishment, and jetties. Property owners are required to follow national guidelines regarding flood and storm insurance, and the state has invested heavily in infrastructure, as well as evacuation and emergency management plans, designed to limit the negative effect of storms.

Ecological modernization is also useful for explaining why restoration has favored technological solutions, even though it is not always clear which technologies will produce better results. For example, beach replenishment, which cost the state more than seven hundred million dollars in the decade before the storm, may or may not have mitigated damages caused by Sandy to oceanfront communities. A related controversy involves oceanfront dunes in the wake of the storm. Governor Chris Christie has been a very vocal supporter of dune construction and reconstruction and has supported the use of eminent domain to wrest oceanfront property from homeowners unwilling to part with their views. Meanwhile, two affluent Monmouth County municipalities debate whether sea walls are a better solution and how sea walls in one community may actually increase dune erosion in another.

Despite the focus on legislative and technical solutions, growing dissatisfaction with governmental actors, scientific experts, and other bureaucracies (such as insurance companies) is predictable. Environmental sociologist

William Freudenberg found that people affected by environmental disasters often come to believe that bureaucratic agents do not have their best interests in mind. Freudenberg called this a sense of "recreancy," but the idea is more clearly expressed by Valerie Gunter and Steven Kroll-Smith, who refer to this as a betrayal. Gunter and Kroll-Smith describe two scenarios that generate feelings of betrayal. First, organizations may be too incompetent or corrupt to actually meet their obligations, which Gunter and Kroll-Smith call "premeditated betrayal." Alternately, organizations may not have the authority or funds to be able to meet their obligations, so sentiments of betrayal belie a lack of understanding between victims and bureaucratic capacity. Gunter and Kroll-Smith refer to this as "structural betrayal." In reality, the failure of bureaucratic agencies may never be conclusively attributed to either premeditated or structural origins, although such failures tend to reduce public confidence in these organizations.

Declining faith in bureaucratic institutions is consistent with Ulrich Beck's "risk society," which is applied to the Jersey Shore in chapter 6. Beck wrote that once a society was able to consistently meet its basic needs, it would cease to organize around material well-being and begin to organize around potential threats. The purpose of government, for example, shifts from a responsibility to provide a social safety net to the responsibility to protect citizens from a variety of real and perceived threats, ranging from economic downturns to food safety to hurricanes. Beck asserts that some risks are shared more democratically than wealth, which is the case for catastrophic threats like climate change. Risk may also invert class position, such as with technological threats like nuclear power, which is concentrated in globally more affluent areas (such as Japan, Europe, and the United States). Risk may be higher for the wealthy and more educated, who not only live in more precarious environments (like oceanfronts) and are exposed to more toxic chemicals; ironically, higher class position is also linked with reduced confidence in scientific or political authorities to mitigate those risks.

Humans have culturally adapted to living with these threats not by relying on experts, but through "habitus": the everyday experience of not needing to face these risks. In describing modern society, sociologist Anthony Giddens explains that we learn to ignore the complex and risky systems on which we depend as a means for preserving our sanity in a world full of potential catastrophes. Giddens suggests that our accumulated experience with nonfailures in these systems produces a sense of "ontological security," allowing us to go about our business without considering risk. When disruptions expose the underlying vulnerabilities of modern society, humans attempt to restore ontological security, even if this does not mitigate the underlying risks. Giddens's observations about complex systems almost perfectly describe the situation that gripped the entire region as a result of the failure of the energy and communication infrastructures in the wake of Sandy. Once the storm had passed, social organization was strained by lack of energy both to homes and to vehicles. Violence broke out at gas stations, requiring gas rationing to be implemented, but people grew increasingly cranky regarding the restoration of the electrical grid, and they particularly demonized energy companies that were seen as not doing everything they could to get power back. The malaise that arose from the Garden State during this time is indicative of the creeping discomfort associated with the knowledge that people rely on systems that they neither control nor fully understand. So while Sandy was hardly a worst-case scenario, it illustrates the precariousness and vulnerability of modern social systems, even while explaining human resistance to tackling the underlying risks and vulnerabilities inherent in coastal development.

By investigating Sandy through the lens of environmental sociology, this book concludes that the way that the Jersey Shore prepared for, was damaged by, and has rebuilt from this storm has been structured by human organization that is patterned, explainable, and predictable. In making these observations, this book seeks not only to increase awareness of the explanatory power of environmental sociology but also to provide suggestions that can improve the way that people, both inside and outside the Garden State, interact with their environments. Since Sandy made landfall in October 2012, New Jersey has been told repeatedly that it is "Stronger Than the Storm." It is true that Sandy did not destroy the Jersey Shore, and it is also true that the Shore is being rebuilt in a way that should "harden" the region against damage from future storms. But these impressive technical fixes fail to account for the fact that not all elements of the Jersey Shore are equally strong, and they do little to change the likelihood that future storms will be stronger, more destructive, and more expensive.

2

The Shore of Memories

There appears to be an unwritten rule that when writing about the contemporary Jersey Shore, one must somehow reference Bruce Springsteen. Born in Long Branch in 1949, Springsteen built his reputation as a songwriter and entertainer while storytelling about the Jersey Shore. His earlier works reflect common experiences of forlorn youth and directly reference the landscapes of Monmouth County; his second full-length album contained a song titled "4th of July, Asbury Park (Sandy)." This is fundamentally a song about a young man hanging out at the boardwalk, and like many of Springsteen's works, centers on trying to hook up with girls and the growing realization that youth is about to end. The object of affection in this particular song is a girl named Sandy, who (like Wendy in Springsteen's much more popular "Born to Run") personifies the singer's hopes and dreams, but is also somehow as trapped as he is in his circumstances. Springsteen could not have foreseen a future where a storm named Sandy unleashed upon the Jersey Shore an upheaval comparable to the transition from youth to adulthood—a period of loss, reflection, and nostalgia, as well as a period of building a future within realistic limits and finding a way to simply move forward in time. Sandy affected not just a place, but the people who have historic, cultural, personal, and emotional ties to that place. To understand the impact of Sandy, then, one must consider how people understood the Shore prior to October 29, 2012.

Geography and Human Ecology of the Shore

The Jersey Shore is both a physical and a social place, and the two roughly but not perfectly correspond to one another. Map 1 provides a basic reference map for the coastal areas of New Jersey. The physical Shore is more clearly delineated, although it is much more varied than is often appreciated. All of the Shore corresponds to the geological feature of the Outer Coastal Plain, the flat landform that was under the sea itself as recently as the last glacial period. The region was forested when Europeans arrived in the region, with mixed hardwoods dominating in the inland and highland areas, cedar in the wooded swamps and barrier islands, and vast pinelands throughout much of the interior lowlands. Even before European contact, this region had been shaped by human presence; the Lenape practiced forest burns for hunting and gathered in large groups along the seashore. Europeans more drastically altered the landscape—a trend that has continued to the current day. The Coastal Research Center at Richard Stockton College (near Atlantic City) reports that only 31.2 miles of the state's 130 miles of ocean beach are undeveloped.[1]

Designating the Jersey Shore is not particularly straightforward. New Jersey's Coastal Area Facility Review Act (CAFRA) defines an area that roughly corresponds to areas east of the Garden State Parkway, with additional buffers along the Raritan and Delaware Bays, as having direct environmental impact on the Shore. CAFRA does not extend far inland, severing the inland regions that are both socially and ecologically connected to the Shore. In contrast, watersheds are among the most important natural divisions, as they link the land, water, and residents through a shared water source. In New Jersey, the four counties that front the ocean, from north to south, are Monmouth, Ocean, Atlantic, and Cape May. These correspond fairly well with four of five major coastal watersheds. The New Jersey Department of Environmental Protection (DEP) has designated five watershed management areas (WMAs) that roughly define large watersheds areas in New Jersey's coastal areas. From north to south, these are:

1. The Monmouth WMA extends from the south bank of the Raritan River along the shoreline of the Raritan Bay, around Sandy Hook and the Navesink-Shrewsbury estuary, and south along the Atlantic coastline to the Manasquan River, and includes most of Monmouth County. This region is characterized by tidal rivers and rich salt marshes and wetlands. Today, this watershed is among the most densely populated nonurban watersheds in the country. Along with Ocean County, this region faced the strongest wind and wave action associated with Sandy.
2. The Barnegat WMA corresponds roughly to Ocean County, stretching just south of the Manasquan Inlet, including both the densely settled Barnegat Peninsula and Long Beach Island. This watershed includes the rivers that empty into the Barnegat Bay, through one of the fastest growing suburban regions in the state. The Barnegat Bay struggles with a variety of environmental problems associated with runoff from its suburban watershed, including fertilizer and pesticide waste. This region was subject to the worst of Sandy's destructive powers.

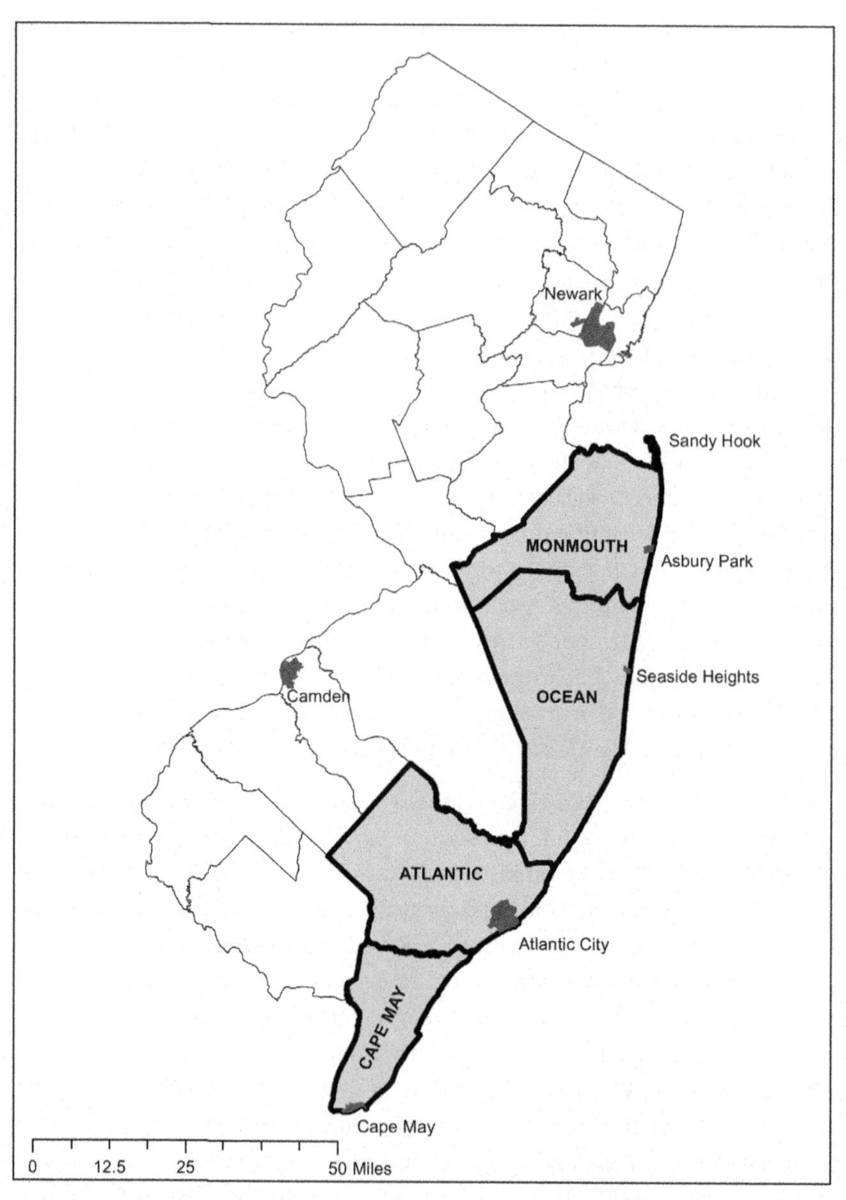

MAP 1. The four counties of the Jersey Shore. Map by author, based on GIS shapefiles from New Jersey Department of Environmental Protection, "NJ-GeoWeb."

3. South of this is the V-shaped Mullica WMA, which includes the Mullica River basin, which empties from inland Burlington County into murky salt marshes and undeveloped barrier islands. This watershed includes the central Pinelands Reserve, where rivers drain intact pine forests such as New Jersey's largest protected area, Wharton State Forest. The Mullica estuary is likewise relatively undeveloped, emptying into the Great Bay surrounded by the Edwin B. Forsyth National Wildlife Reserve. Relatively little of this watershed is contained within the four coastal counties. As a consequence and because the waterfront areas in this region are relatively undeveloped, Sandy's impact was mainly inland and related to wind.
4. Farther south is the Great Egg Harbor WMA, which flows into the Great Egg Harbor River and includes much of Atlantic County. Draining the southern Pinelands, the Great Egg Harbor WMA also includes the densely developed barrier islands containing Atlantic City and Ocean City and the expanding sprawl of the Atlantic City Expressway Corridor. The eye of Sandy made landfall in this watershed, so there were substantial impacts by wind and water, although they were not as destructive as farther north.
5. Cape May WMA is the southernmost area, which more or less corresponds to Cape May County, including some waterfront on the Delaware Bay. This watershed contains densely settled barrier islands, such as Sea Isle City, dense mainland coastal areas on the Cape May Peninsula, extensive salt marshes, and inland residential and agricultural areas. Sandy dropped

2.1. Sea Bright Bridge and Navesink Estuary from above, fall 2013. Photo by author.

2.2. Manasquan River Inlet separating Manasquan and Point Pleasant Beach from above, fall 2013. Photo by author.

> more rain in this region than elsewhere in the state, but because this region was south of the eye at landfall, the wind, surge, and wave action were not as severe as in other coastal regions.

The diversity of these watersheds demonstrates complex interactions between humans and their physical environment, and these are overlaid and crossed with social and political divisions. Social and environmental analysis of the Jersey Shore is facilitated because watersheds more or less correspond to political counties, a common unit of analysis for social and economic studies; counties may also have distinct cultural attributes. For example, Monmouth and Ocean Counties have stronger economic and cultural ties to the New York metropolitan area, while Atlantic and Cape May Counties are linked more closely to Philadelphia. Long Beach Island, the southernmost barrier island in Ocean County, is caught in the middle, enjoying its reputation as a tranquil escape from the livelier Shore towns to the north and south.

Crosscutting these watershed and political divisions, the Shore can be divided into five overlapping human ecological subregions with somewhat unique current and historic relationships between people and the physical landscape. The combination of physical features and social uses explains why Sandy damaged each sub-region in predictable but somewhat different ways.

First, there is the Raritan Bay shore, which stretches from the mouth of the Raritan River at Sayreville and runs east to Sandy Hook, including the Atlantic Highlands. This region is noted for its species-rich estuaries and proximity to New York Bay, which combine to make it a popular location for people to find consumable seafood. However, since 1982, New Jersey's Department of Environmental Protection has advised against consuming quantities of fish and shellfish caught in the Raritan Bay because of high levels of bioaccumulated carcinogens including polychlorinated biphenyls (PCBs), dioxins, and heavy metals.[2] Predominantly white, the Raritan Bay shore has a fairly working-class population and identity, although this has changed somewhat through gentrification since the introduction of direct ferry service to Manhattan in the 1990s. No longer dependent on fisheries, residents of the Bayshore continue to be disproportionately concentrated in manual trades, with lower levels of formal education and income than elsewhere in the region and in New Jersey as a whole. Because of the shape of the Raritan and Lower New York Bays, this subregion experienced the highest storm surge during Sandy and consequently suffered some of the most extensive damage associated with the storm. One of the iconic photographs of the storm, which pictured the remnants of a modest home with its lower left side shorn away by water, was taken at Union Beach on the Raritan Bay shore.

Second, there are the headlands from Monmouth Beach to Point Pleasant Beach in Monmouth County.[3] Geologists refer to this landform as a headland because the ocean directly erodes a low bluff of the continental shelf. Locally, these are often referred to as "mainland" beaches, and these are among the most accessible beaches on the Atlantic coast of the United States. They are also home to some of the oldest year-round beach communities on the Jersey Shore. Mainland beach communities have been drastically altered from their natural condition. Land developers filled in tidal swamps and lakes to create waterfront property, although other lakes continue to exist in an engineered form, including Wesley, Deal, Fletcher, and Sylvan Lakes, as well as Lake Como and Spring Lake. The Monmouth headlands beach towns are among the oldest and wealthiest beachfront communities in the United States, although some resorts like Long Branch and Asbury Park have seen periods of decline, neglect, and redevelopment. This is a densely developed and eroding coastline, but it is considerably more affluent than the neighboring Bayshore and inland towns, and as such, has greater resources on which to draw for both mitigation and redevelopment. During Sandy, the headlands suffered beach and bluff erosion as a result of storm surge and wave action that destroyed buildings and boardwalks. Moreover, the ocean forced its way back into many of the "lakes," often through narrow engineered channels, causing extensive lakefront flooding.

A third human ecological region includes the barrier features, which stretch in an almost continuous line from Bay Head to Cape May. Sea Bright, technically

2.3. Beach at Asbury Park, summer 2013. Photo by author.

an engineered spit that separates the Shrewsbury and Navesink estuaries from the Atlantic and connects Sandy Hook to the mainland at Monmouth Beach, also has many of the characteristics of a barrier island. The "island" stretching between Bay Head to Island Beach State Park is attached to the mainland, and is actually called the Barnegat Peninsula, but it maintains the form and function of a barrier island and is often locally referred to as such. In their natural state, barrier islands are permanent but somewhat ephemeral landforms that separate bays from the open ocean. The typical profile of a barrier island would begin with an oceanfront sand beach, rise up to increasingly vegetation-stabilized dunes, then slope downward through scrub forests and a marshy leeward shoreline and eventually back to a salt or brackish water bay. The ocean beach and initial line of sand dunes present natural water and wind barriers to the leeward side of the island and the bays behind them. Barrier islands can be fairly broad and give the appearance of dry land, but because they are composed mainly of sand and muck, they have historically confounded permanent development. Even so, New Jersey's barrier islands are among the most modified and developed in the world. With the exception of Island Beach State Park and the Forsythe National Wildlife Refuge, the barrier islands are densely developed and essentially built out. The explosion of homes and second homes on barrier islands after 1945 created strong incentives to stabilize the shifting sands regardless of the environmental wisdom of doing so. Barrier islands communities have a unique social dynamic, with huge swings in seasonal populations, and a relatively modest year-round population (in terms of both numbers and socioeconomic status) and a much larger and more affluent crowd in the summer. The barrier islands north of Little Egg Harbor were devastated by Sandy's surge and wave action. The New Jersey Department of Community Affairs (DCA) estimates that in some barrier island communities up to 85 and 90 percent of residences suffered "major or severe" damage during the storm, with clusters of acute damage near Mantoloking, Ortley Beach (Toms River Township), and central Long Beach Island.[4] Barrier islands also experienced flooding from the bayside, as storm surges pushed water up into the marshes and estuaries.

Landward from the barrier islands lie the bay communities that front protected waters from Brick Township at the northern end of Barnegat Bay to Cape May's Lower Township and the towns located along tidal rivers, including (from north to south) the Navesink, Shrewsbury, Shark, Manasquan, Toms, Mullica, and Great Egg Harbor Rivers. Like oceanfront communities, bay and river towns have drastically altered their natural state. The U.S. Environmental Protection Agency estimates that 36 percent of Barnegat Bay's shoreline has been bulkheaded,[5] completely eliminating the natural boundary between water and land. Like the barrier islands, much of the development in the bay and river communities occurred in the postwar period, transforming sleepy fishing and farming villages into densely settled and increasingly affluent year-round residential communities. Many bayside communities feature "lagoons," which are artificial waterways that provide additional bulkhead access to natural waterways. Bay and river towns were mainly affected by high winds and floods caused by the ocean surging up rivers and into bays and preventing the normal flow of the regions rivers. Unable to disgorge their normal flows, rivers backed up rapidly into their floodplains, much of which had been converted to human uses. The same process backed the bays up into waterfront communities, far from the actual ocean.

The final human ecological region of the Jersey Shore involves the inland communities that are closely linked to the Shore economically, culturally, politically, and environmentally, but have no salt or brackish waterfronts. In almost all cases, these are communities that are ecologically tied to the shore through watersheds. They are expansive areas in the Shore's interior, often associated with the Pinelands, but essentially a suburban and exurban corridor running along the Garden State Parkway from the Raritan River to Toms River; south from there, development tapers to Long Beach Island and then dissolves into patchwork development in the marshes of Atlantic and Cape May Counties, including a corridor along the Atlantic City Expressway. Of the subregions, the inland regions of coastal counties have grown most dramatically in the postwar period. The conversion of vast swaths of the Pinelands and agricultural fields to subdivisions catering to a variety of social and income groups is arguably the most distinguishing environmental transformations in the state over the past half century. Like suburban areas elsewhere, the inland Shore is notable for its tree cover; as a consequence, Sandy affected this region primarily by disrupting the energy grids as trees systematically knocked down power lines. The inland Shore also became the primary refuge for people fleeing waterfront communities and served as a staging area for disaster recovery

Many Jersey Shore municipalities contain elements of all five of these human ecological subregions. For example, Manasquan is a headlands beach town, but it suffered extensive damage when the Manasquan River overflowed its banks. Larger communities in Ocean County, like Toms River, Brick, and Berkeley Township, have part of their land on the barrier island, part of their land fronting the Barnegat Bay and along coastal rivers, and even more land stretching inland into the Pinelands. Residents of all five subregions may identify themselves as living "down the Shore," although people frequently distinguish between inland and waterfront communities by indicating "the beach," or "the bay," or "the island."

A Brief Environmental History of the Jersey Shore

How did these subregions develop their distinct social and environmental characteristics? Cultural sociologists emphasize the importance of understanding a region's environmental history to understand emotional and normative connections to place. To this end, it is important to understand how humans came to interact with the Jersey Shore and its environs.

Prior to the European conquest, small groups of the Lenape (Delaware) lived on what would become the Jersey Shore, and nonresident Lenape had large seasonal gatherings in the coastal areas, leaving immense shell middens from Keyport to Tuckerton.[6] Compared to the Delaware River Valley, relatively few Lenape lived in the Shore region, owing in part to poor agricultural land in the Outer Coastal Plain and better transportation and denser social networks along the Delaware River. The Shore area of what is now New Jersey was divided among two major divisions of the Lenape, with the Unami (sometimes referred to as the "Turtle Delaware") occupying most of Monmouth and Ocean Counties, and the Unalachtigo in southern Atlantic and Cape May Counties, as well as the Delaware Bay shore.[7] Because the Lenape shared a language and had generally peaceful relations with each other, trade and intermarriage between the two groups was common.[8] Lenape economic activity centered on the cultivation of corn, but larger groups gathered seasonally to take advantage of fish and shellfish resources along the coast. Anthropologists suggest three basic resources that shaped how the Lenape used the Shore region: seasonal fish runs in coastal rivers; shellfish in tidewaters and bays; and upland wildlife, including deer and fruiting plants.[9] Lenape made use of hollowed-log boats to visit islands along the coast, but there is little evidence to suggest that they made durable settlements on the barrier islands.

The first European record of the Jersey Shore was in September 1609, when Henry Hudson and his ship *Half Moon* entered the Delaware Bay, then traveled up the coast, landing near Great Egg Harbor. Hudson then traveled north, describing the Barnegat Inlet and anchoring "within sight of the Highlands."[10] He sailed north and west into the Raritan Bay, anchored off Keansburg, and sent a party ashore, where the ship's log noted forests of "great and tall oaks" and rich fish resources.[11] The crew of the *Half Moon* spent a few nights in the area, trading with the local Navesink Lenape, whom the ship's log describes as "very civil."[12] Hudson's log describes a "great store of men, women, and children" in the area. Trade included cultivated tobacco, hemp, and maize (including "good bread") as well as fresh and dried berries. The Navesink wore deerskins, feather mantles, and furs, and had copper ornaments and pipes. Hudson sent a smaller party north into Newark Bay, where they were "set upon" by natives in two large canoes (carrying twelve and fourteen men), who killed an English sailor, John Colman. Colman was later buried on the mainland; it is suggested that his burial site is in the region known as Colman's Point in Keansburg. Although the local Navesink exhibited no hostility toward the *Half Moon* thereafter, Hudson's crew was leery and left the Jersey coast within three days.

Patterns of European settlements generally followed that of the Lenape. Europeans made use of Lenape trails and settlement sites, even while gradually eliminating the native populations through land purchases, disease, violence, and, eventually, forced displacement. Dutch traders recognized the superior natural harbor of the New York Bay and encouraged natives to exploit the region's fur resources to trade for European technologies. The Dutch generally failed to settle the region, except along the Navesink and Shrewsbury Rivers. The governor of New York offered a patent for much of Monmouth County to English settlers from Long Island in 1665. Two years later, in 1667, Dutch officers reporting on loyalty oaths from colonists in the region noted the presence of sixty men in Middletown and sixty-eight in Shrewsbury,[13] indicating fairly substantial settlements, although no mention is made of the size of the populations of these communities as a whole. In the late seventeenth century, England permanently took control of Dutch-controlled territories including New York and New Jersey. The English divided the state into colonies of East and West Jersey along a diagonal line that stretched to the northwest from Little Egg Harbor, thus reinforcing the north-south division of the coast that roughly corresponds to both contemporary and the Lenape cultural divisions. English immigrants quickly settled in fishing villages, and they planted farms along the Raritan Bay and the Shrewsbury River, as well as developing smaller communities on rivers farther to the south. The sandy soils, poor transportation connections to population centers, and lack of natural harbors left most of the rest of the Jersey Shore a sparsely settled backwater region—a situation that persisted well into the twentieth century.

One exception to this involves the early development of health resorts along the Jersey Shore. Both Long Branch and Cape May claim to be the first seaside resort community in the United States, with evidence suggesting that both attracted tourists as early as the late 1700s.[14] During the next century, many other oceanfront communities developed facilities to attract and service visitors seeking escape from the oppressive heat of New York and Philadelphia summers.

The heyday of Jersey Shore resorts was probably in the 1920s. The boardwalks became true destinations in Prohibition era, not least because New Jersey's long coastline provided opportunity for importing bootlegged liquor and the state's readily corruptible political system tolerated the organized crime involved in the trade.[15] Given the region's steamy summers and the Shore's close proximity to major metropolitan areas, seaside resorts were a natural destination for tens of thousands. This was the most glamorous period in the Shore's history, when presidents, celebrities, and gangsters all mixed and mingled with everyday (white) people.

Historian Bryant Simon argues that because the Atlantic City Boardwalk was a promenade located in a resort area, people were freed of the constraints that come from displaying status among people who know you in your everyday life.[16] Social status at a resort was determined wholly by how you behaved, what you wore, and your leisure activities, rather than by the more mundane social rankings of occupation, income, and ethnicity. To Simon, the Atlantic City Boardwalk ushered in a new phase of tourism whereby vacations allowed the middle classes to enjoy the lifestyle of the wealthy, if only for a night, or a weekend, or a week. Vacationers were also attracted to the more exotic attractions that Atlantic City offered, including illegal speakeasies, burlesque shows, and African American musicians. As these attractions became more prominent, more affluent and family-oriented visitors began to consider other locations for their holidays on the Jersey Shore.

Vacation home communities developed in conjunction with resorts, but continued to grow even as the larger resorts began to decline after World War II. Working- and middle-class families in North Jersey and Philadelphia invested savings into modest homes, often pooling resources across extended families. A "Shore house" does not necessarily refer to a large, posh property overlooking the ocean; the more traditional beach house is a small ranch or Cape Cod shoehorned into a small lot, surrounded by other similarly small and modest houses. By the turn of the twenty-first century, second homes and rentals became more typical of visitor experiences than extended overnight stays in resort hotels, especially in Monmouth and Ocean Counties. Boardwalks and amusement centers continue to attract visitors, but the draw shifted to thrill rides and arcades and away from promenading and status enhancement. Some amusement centers deteriorated and were eventually ceded socially to more marginal groups in society, or narrowed their focus to family amusements, sometimes specifically excluding alcohol sales and nightclubs. Although not all of the Shore is a playground and not all people "play" the same way at the Shore, people continue to think about and define the Shore as a place of recreation, not routine activities.

The Shore in Popular Culture

Bruce Springsteen looms large in the cultural iconography of the Shore that emerged in the modern era. His first two albums, both released in 1973, were filled with quixotic and willful young people, coming from a working-class background ("cages on Highway 9") but dreaming of better things. This is the imagery and the soundtrack for a generation that came of age in the era of deteriorating seaside resorts. By the time Springsteen lionized the Shore, its boardwalks were fading from their past glory in favor of shiny new suburban shopping centers and theme parks. Disneyland opened in 1955 in Anaheim, California, and fundamentally reshaped the ideal middle-class family vacation. Suburban homes filled with televisions and air conditioning reduced the lure of ocean breezes, and Shore resorts were never able to recapture their desirability among vacationers (more on this in chapter 3). People continued to spend time at the Shore, but the shift from long stays in resorts with amusement centers like Atlantic City and Asbury Park to second homes and day tripping via the Garden State Parkway (completed in 1956) and the Atlantic City Expressway (completed in 1965) did not reverse. Springsteen used nostalgia and decay to artistic effect, and his vision of the Jersey Shore resonated with baby boomers nationwide who had come of age in the shadow of the Greatest Generation, with humble roots but big dreams.

Springsteen also documented the bleak impacts of deindustrialization and the decentering of the blue-collar worker in the American Dream. Although the Shore was never really industrial, there was enough industry in the area, particularly in adjacent inland towns and the Raritan River Valley, that the decline of Shore resorts is, in Springsteen's lyrics and in the popular imagination, linked to the larger disappearance of factory work. Deindustrialization marked the end of the era for blue-collar social mobility, as factory jobs moved away from states like New Jersey, where labor costs were relatively high. For the white working class, which had enjoyed the dramatic expansion of its purchasing power and high material standards of living in the postwar years, the late

1960s and early 1970s were a period of relative loss. This melancholy resonates in Springsteen's presentation of the Shore, which employed both the crumbling imagery of the physical environment and the anomic angst of the white working class. The Jersey Shore that Springsteen offered to the national and regional imagination was full of nostalgic sentiment and romantic notions of escape, set against a backdrop of bleak decay.

Visitors to the old Shore resorts like Atlantic City and Asbury Park in the 1970s undoubtedly found landscapes as cheerless as Springsteen described. Asbury Park experienced its own riots in 1970, sparked by a dance-hall fight and fueled by a history of racial antagonism and police brutality. The subsequent decade saw an almost complete abandonment of the city by middle-class residents and tourists; the boardwalk amusements were literally abandoned and then fell apart. Atlantic City was able to jockey its failing summer tourism industry into the legalization of casino gambling in 1978. But visitors to the "Queen of Resorts" found decrepit housing, abandoned properties, and lots and lots of homeless people outside the casinos. Springsteen's 1982 album *Nebraska* featured the song "Atlantic City," a folksy homage to the grim hope of gamblers that a win can make all their problems go away. It was paired with a black-and-white video that featured the demolition of landmark hotels, the tacky functionality of the earliest casinos, and the armada of unglamorous buses that brought gamblers to their fate. The song and the video reflect the Shore as Springsteen fans imagined it: partially demolished and desperately hopeful. By 1984, the Boss became much bigger than the Shore, with the release of his multiplatinum *Born in the USA*, which was less rooted in New Jersey's cultural geography.

That same year saw the first release of a scrappy rock band from Sayreville, led by a charismatic frontman who adopted the stage name Jon Bon Jovi. With three of the original members born in Perth Amboy and another from Colonia, Bon Jovi's early history is more closely tied to the Raritan River Valley than the Shore exactly, but this hair metal group has had a long association with the Shore. Bon Jovi's music rarely spoke to the specifics of the Shore's landscape beyond generalizable references to waterfronts and oceans ("Tommy used to work on the docks," in "Living on a Prayer"), but after becoming the best-selling musical act in 1988, it followed up with an album simply titled *New Jersey* that redirected Springsteen's nostalgia about the Shore to a self-indulgent hedonism that was evidently proud of its place of origin but functionally independent of place. In this way, Bon Jovi captured the Shore of the 1980s, one that had not given up on the American Dream, but had abandoned poetic melancholia in favor of an in-your-face party attitude. Despite its surprising lack of references to the Jersey Shore, Bon Jovi has had a lasting impact on the region's image in popular culture, with its focus on the disaffected white working class and the self-centered pursuits of young people who could literally be living anywhere.

This portrayal of postindustrial wastelands was revisited and popularized by a series of movies written and directed by Kevin Smith, who rejected Bon Jovi's hedonism in favor of snarky nihilism. *Clerks* (1994), filmed in Leonardo, portrays the dead-end lives of two twenty-somethings on New Jersey's Bayshore who do little more than talk (mostly about girls, sex, science fiction, and comic books), work at mindless jobs, and make some minor mayhem. The store they work in is visited by a cast of characters including Silent Bob, a philosophic burnout (played by Smith himself), Jay, a wisecracking pot dealer, and a surreal parade of people that the eponymous clerks knew in high school but who had moved on to bigger and better things. Famously made on a shoestring budget, Smith's independent film hit a chord with a national audience, which saw itself as more cerebral than the Bon Jovi fan base but nonetheless continued to pigeonhole the Jersey Shore, particularly the Bayshore, as a place where the white working class existed, and was getting by, but was not really going anywhere. Jay and Silent Bob went on to be featured in five more movies that continually reference Shore locations, as well as a short-lived animated series. Jay and Silent Bob also lend their names to the comic book store that Smith owns in his now-gentrified hometown of Red Bank, which now serves as the locale for a reality show about comic book fans.

The most recent pop cultural send-up for the Shore comes from MTV's reality show, *Jersey Shore*, which documented the hijinks of eight young people sharing a summer rental in Seaside Heights. The show, which gained rapid popularity and notoriety, featured the shallow, bacchanalian daily life of a cast of caricatures, only one of whom was from New Jersey (although another Jersey girl did join the cast in the third season). *Jersey Shore* updated the national imagination of the actual New Jersey Shore by offering up negative stereotypes of white ethnics as garish hedonists in line with Bon Jovi's portrayal nearly a generation earlier. The personalities featured on the show spent most of their time engaged in self-absorbed activities, including working out, hanging out at the beach, drinking, dancing, and hooking up with other people doing the same thing. The show played to, and was widely criticized for, negative stereotypes of Italian Americans, using the generally derogatory terms "Guidos" and "Guidettes." The show undoubtedly raised the profile of Seaside Heights as a party town, albeit for a rather unsophisticated crowd, but locals quickly began to wonder if its new popularity was a good or a bad thing. In 2011, Governor Christie canceled a nearly half-million-dollar tax incentive for production of the show, explaining, "I am duty-bound to ensure that the taxpayers are not footing a $420,000 bill for a project which does nothing more than perpetuate misconceptions about the state and its citizens."[17] After Sandy, the residential community of Pelican Island, just west of Seaside Heights, blocked a *Jersey Shore* spinoff, with concerns about privacy and interference with restoration from the storm.

In sum, the Jersey Shore in modern popular culture has endured some persistent themes that have affected the way the entire country (and probably the entire world) views the region. First, there is a persistent belief that it is peopled with directionless working-class whites who are alternately introspective, hedonistic, caustic, or shameless. The Jersey Shore thus serves as a stark contrast to its near neighbor to the north, Manhattan, with its cultural sophistication, multicultural and global sensibilities, and dazzling wealth. In this way, the Jersey Shore has become a cultural straw man and easy target for cultural memes. This is not to say that there are no truth in these cultural representations, but that that these stereotypes obfuscate a much more complicated region.[18]

The Shore in the Local Culture

At the turn of the twenty-first century, the national imagination of New Jersey was dominated by HBO's *The Sopranos*, which chronicled a contemporary Mafia boss and his suburban family. Debuting in 1998 and taking place mainly in North Jersey, *The Sopranos* did not use the Shore as a plot device until the final episode of its fourth season, when the series protagonist (or antihero) sought to buy a house in Sea Bright in a desperate attempt to glue his increasingly estranged family together. Episode writers David Chase, Robin Green, and Mitchell Burgess drew on many of the important cultural associations with those Shore towns known more for their second homes than for boardwalk amusements. Shore homes, although rarely as extravagant as the one depicted in *The Sopranos*, have been important status symbols for New Jersey's upwardly mobile middle classes. This was an important departure for a representation of the Shore to a national audience, but despite the critical acclaim for the episode, which earned the writers an Emmy in 2003 for Outstanding Writing in a Drama Series, it is likely that the symbolism of the Shore house resonated most strongly among viewers with firsthand knowledge of the Shore.

For New Jersey residents and for people in nearby states, who together constitute roughly 80 percent of the state's tourists, the cultural importance of the Shore is quite different from how it has been portrayed in national cultural scripts. The Shore is more about family-oriented leisure, childhood, and recreation than about disaffection and over-the-top revelry. And although it is not the playground of the ultrawealthy, neither is it entirely a region of down-and-out whites. The Jersey Shore is remarkably democratic in terms of being accessible to and actually enjoyed by a wide spectrum of classes, races, and ethnicities. This is not to suggest that there is not substantial segregation and outright exclusion, but there are few other ocean shorelines in this country that support such a diverse group of residents and visitors. And while the nightlife in some areas has grown in popularity and rowdiness in the last two decades, most of the

Shore remains deliberately quiet and sober, catering more to families than to visitors focused on debauchery. Governor Chris Christie addressed this directly when he spoke to crowds at the end of the summer 2013 season in Point Pleasant Beach:

> There's nothing more Jersey than the Jersey Shore. Nothing more Jersey than our boardwalks. The first boardwalk in the country was built right here in New Jersey down in Atlantic City in 1870, and ever since then boardwalks have been being built. Twenty-four beautiful lighthouses up and down our coast. And even seven presidents have vacationed on the Jersey Shore. The Jersey Shore has been a place for families more than anything else though, more than presidents, more than governors, even more than Snooki and The Situation. The Jersey Shore has been about families coming here and making memories with their parents and grandparents, with their children and grandchildren, with aunts and uncles. And so, when I came down here in the immediate aftermath of the storm and saw the destruction here in Point Pleasant Beach, it broke my heart.[19]

This depiction of the Shore as a place where families take their vacations has been deliberately promoted by the New Jersey Division of Travel and Tourism. Its 2012 visitor's guide cover features a young girl running up the beach with a kite in her hand and the ocean at her back. The girl's mother is smiling at her from a sand chair, with a book open in front of her and a wicker picnic basket at her side. Behind her, closer to the ocean, a young couple rests on a blanket under an umbrella, watching what appears to be another family playing football just above the waterline. After five pages of advertisements, Governor Christie invites potential visitors to make of use of the guide in order to "organize your next *family* trip, whether it's a quick adventure or a weekend excursion" (emphasis added). The first real content (after a paid advertising section about Newark) is a brief article about New Jersey's boardwalks, which describes Point Pleasant Beach as "super family-friendly"; Seaside Heights as a "magnet for young people, featuring game stands, rides, arcades and even a water park"; and Spring Lake as "unhurried and peaceful." The official "Shore Region" encompasses Monmouth and Ocean Counties (Atlantic and Cape May Counties are covered in separate sections), and it is the first of the guide's regional profiles. Among the nineteen points of interest in the "Shore Region," six explicitly mention "family," "children," or "all ages" and another two are clearly oriented toward families, including the Point Pleasant Beach boardwalk and an indoor amusement park in Freehold. In comparison, only two of fourteen points of interest in New Jersey's other main tourism region, the Skylands of northwestern New Jersey, mention families, children, or all-ages activities; only two of thirteen points of interest in the populated Gateway region in the northeastern part of the state do so.

Although the boardwalks of today are not the same as they were in the heyday of the large resorts, the region's visitors (and targeted visitors) remain firmly rooted in family-oriented, middle-class leisure activities. Thus, despite the widespread stereotypes mentioned in this chapter, the strongest cultural associations for people who actually know the Shore are overwhelmingly positive and personal. Sandy's destruction damaged these memories; Bay Head Councilwoman D'Arcy Margaret Rohan Greene explained the grief she encountered from second-home owners in her town who suffered severe losses: "That was their home, as young people, as young children. That's where their summer memories are from."

Environmental Claims and Double Realities

A storm like Sandy does not affect just a physical location. The storm struck an environment heavily modified by human activities and imbued with human meanings. Human alterations to the natural environment help explain why there was so much damage in the storm—if there were no permanent structures on the beaches or in the inland floodplains, there would have been very little damage at all. But the response of people to Sandy's destruction is a reflection not just of the damaged human environment but of what those environments mean to the people who have lived and played there. To understand the Shore, it is important to understand what the Shore actually is, both in the "real" world and in the collective imagination. How people imagine the Shore, or as sociologists might say, how they "construct" the Shore, is as important as the physical reality, because the way that people think and feel affects their decisions regarding their use of the physical space.

Sociologist John Hannigan published *Environmental Sociology: A Constructionist Perspective* in 1995, the most long-standing theoretical contribution to environmental sociology from a constructionist tradition. Hannigan revised and expanded on these ideas ten years later in *Environmental Sociology* (2006, deliberately with no subtitle in the second edition).[20] Hannigan explains that how people come to think about environmental problems depends only partially on scientific evidence; more important is how such issues are framed to resonate with local experiences, linked to existing moral justifications for action, and marketed to receptive audiences. Post-Sandy reconstruction thus must be tied to how people think about the Shore in their personal experiences, must motivate people to act in a way that is morally consistent with those experiences, and must be more broadly packaged to generate support. To this end, how people thought about and continue to think about the Jersey Shore prior to Sandy allows us to make sense of the call to "Restore the Shore," even in light of widely accepted scientific evidence that future destruction is almost inevitable.

Hannigan's analysis begins with the idea that social problems, or, more broadly, social concerns, are rooted in "claims" that are constructed by actors who make observations ("complaints") about them in order to mobilize people to act. Claims are rhetorical attempts to persuade others to a specific way of viewing a social problem in order to resolve it in a way that suits the needs of people asserting claims. As such, claims rely on a variety of persuasive techniques, which Hannigan calls grounds, warrants, and conclusions. Grounds include empirical data used to convince others that a problem exists; Hannigan notes that in addition to scientific data, "practical" or "ordinary" knowledge may be used as grounds. Warrants involve demands for action based on moral imperatives, such as "presenting the victim as blameless or innocent, emphasizing links with the historic past or linking the claims to basic rights and freedoms."[21] Conclusions indicate appropriate actions for mobilizing around the claim, usually involving institutional intervention and/or social policy. Hannigan also emphasizes that some claimants having greater authority and persuasive abilities than others, so it is also important to consider who is making which claims.

Claims must first be "assembled" or identified and defined. Environmental disasters like Sandy rarely have to be identified, because they effectively impose themselves on human actors, regardless of any previously held ideas about storms. However, the assembly process is not just about identification, but about definition. To this end, Hannigan specifies that the creation of an experiential focus is critical to assembling a successful environmental claim, and this experiential focus relies heavily on the ordinary, day-to-day experiences of nonexperts interacting with the environment. As noted, since the turn of the last century, the most common association with the Jersey Shore has been one of leisure and recreation, as opposed to productive activities or scientific knowledge. Although storms are a regular occurrence and form part of the region's uncontested history, these environmental threats compete with mundane memories of going to the beach and the boardwalk, typically colored with emotional links to family and friends. Storms like Sandy are intruders in an otherwise continuous summer of memories; restoration becomes about restoring the landscape of these memories. Other potential claims, such as questioning waterfront development or climate change, become secondary. By assembling or defining an environmental claim that resonates with the lived experiences of social actors, presentations of the claim are more easily adopted by people in a position to motivate and participate in social action.

In so doing, claimants must "command attention and . . . legitimate their claim" to a broader audience. Commanding attention relies heavily on rhetorical strategies well known to marketers and public relations professionals, especially the reliance on "evocative verbal and visual imagery."[22] In this way, the three iconic images of Sandy can be used to command attention from people who have not been directly affected by the storm. It is impossible to be unsympathetic to the residents of Union Beach after seeing their houses torn from their foundations, not to be nostalgic upon seeing the Star Jet roller coaster lying in the surf, or not to be astonished by the power of the ocean to carve out a new inlet at Mantoloking. Verbal imagery is equally compelling; residents throughout New Jersey are told they are "Jersey Strong" and "Stronger than the Storm," and thus capable of "Restoring the Shore." Of course, the persuasiveness of the claim is enhanced when it is made by charismatic cultural icons and popular political leaders, including Bruce Springsteen, Jon Bon Jovi, and Governor Chris Christie, all of whom have lent their public images and efforts to restoration, such as in the 12.12.12 concert and countless smaller events since. The claim is also strengthened because it is aligned with the stronger moral frames about the importance of spending leisure time with family and the sanctity of middle-class home ownership.

The final part of constructing environmental problems involves what Hannigan calls "contesting"[23] but is better understood as the social actions taken in response to the environmental claim. This involves direct mobilization, such as volunteer efforts in the immediate post-Sandy period, as well as public actions and policies such as direct aid, redevelopment of infrastructure, and overall planning for reconstruction. If an environmental claim is successfully assembled and presented, action becomes a moral imperative; it would thus be culturally "wrong" not to restore the Shore. Failure to restore is less about the likelihood of future natural disasters than about the elimination of cultural continuity among generations. Restoring the Shore is not just about place or environment but about what makes New Jersey a unique and valuable place where families live and have fun. From this perspective, people who question why the Shore is being rebuilt can thus be disregarded—for they are people who do not share the same cultural understandings of the place, nor do they have the same moral purpose as those dedicated to restoration.

At the same time, the contrast between the "Restore the Shore" mobilization and the widespread and widely accepted knowledge that dense development in coastal regions will inevitably result in destruction by future storms is consistent with what University of Oregon sociologist Kari Norgaard calls a "double reality."[24] In her monograph *Living in Denial: Climate Change, Emotions, and Everyday Life* Norgaard found that Norwegians failed to integrate ample evidence of climate change into their everyday lives, despite having strong levels of cultural identification with their natural environment. As a result, they were unable to devise realistic solutions to reverse or even mitigate a pending local crisis brought on by warmer and drier winters. In her view, climate change was inconsistent with Norwegian cultural identity (which strongly included a connection to nature), beyond the control and scope of everyday living, and attributable to other actors, so locals created normative mechanisms that allowed them to ignore the environmental change that fundamentally threatened their livelihoods and lifestyles. In much the same way, Jersey Shore residents and visitors can ignore the precariousness of waterfront development, allowing and promoting restoration with procedures such as elevation of homes and technologies such as sea wall construction that may mitigate but can never fully protect people and property from future storms. Even in the absence of ongoing concerns about sea-level rise associated with climate change, post-Sandy redevelopment stands as an example of the power of culture over science in how people make decisions about their interaction with the environment.

Norgaard suggests that this double reality is not a denial of risks or a lack of knowledge about risks, but a cognitive adaptation to a reality that is inconsistent with shared meanings about the environment. She chose as her case study a village in Norway specifically because Norwegians are highly educated, including knowledge of climate-change science. Few in Norway doubt that climate change is already under way, in contrast with the United States, where a persistent and vocal minority continues to deny climate change science. Norgaard herself was struck by locals who expressed serious concerns about climate change, but failed to connect those concerns to changes in the local environment, notably a shortened ski season and lack of snowfall. Through ethnographic research and interviews, Norgaard eventually discerned a "socially organized denial" to explain apathy in the face of uncontested scientific knowledge. She calls this socially organized because the denial is not unique to individuals, but normative throughout the community.

In much the same way, restoration of the Jersey Shore is done with widespread acceptance of future storms and climate-related sea-level rise. A Rutgers-Eagleton survey conducted in April 2013 found that 65 percent of residents thought that Sandy and other severe weather events in the previous two years were a result of climate change. Nearly half (49 percent) indicated that it was "very" or "somewhat" likely that climate change would cause a "natural disaster" *in their home community within the next year* (emphasis added).[25] One year after the storm, the state's most widely read newspaper, the *Star-Ledger*, dramatically and in great detail criticized the lack of attention to climate change and sea-level rise in the reconstruction effort, citing both policy in other states affected by Sandy and several Rutgers University scientists.[26] Unlike Norway, New Jersey had its own set of climate-science doubters (including the governor himself), but only a minority of New Jersey residents (the Rutgers-Eagleton survey estimates about 29 percent) see no connection between climate change, Sandy's destruction, and future devastation from storms. This makes denial of climate-change science an unlikely motivation behind the intransigence of individuals and entire municipalities about pursuing reconstruction designed to mitigate against such damage, including increases in elevation levels, which are being instituted in New York, Delaware, and Maryland. The call to "Restore the Shore" thus exemplifies people's ability to acknowledge risk but fail to pursue social actions designed to respond to that risk.

Although it was not a focus of her research, Norgaard also pointed out that reliance on North Sea petroleum gave Norwegians a vested interest in not dwelling too much on how their own behaviors fail to protect them from climate change. In much the same way, the Jersey Shore and the entire Garden State are so dependent on revenues generated from tourism and vacation real estate that failure to restore could work against the region's economic interests. It is to this issue that Chapter 3 turns, investigating not how people think about the Shore and its reconstruction after Sandy, but the economic imperatives of rebuilding.

3

Shore Resorts

Just as it is impossible to ignore Bruce Springsteen when considering Jersey Shore culture, it is difficult to introduce the development of the Shore's tourism industry without reference to Monopoly. Milton Bradley's popular board game is based on real estate acquisition and development, and its familiar streets are named after real counterparts in Atlantic City. The game's goal is to force competitors into bankruptcy while amassing an empire of hotels on differentially valuable proprieties, including a park and the seaside boardwalk. This goal coincides with but does not fully reflect the real-world activities of real estate developers on the Jersey Shore, where the most valuable properties are those in proximity to natural amenities—that is, the waterfront. What both the game and the region's economic boosters have obfuscated is how real estate interests coalesce into alliances that defy natural limits, exclude large segments of the population, and dominate the political system to funnel public resources to private gains.

The Jersey Shore's environment was intentionally developed as a real estate venture, with capitalist investors promoting resort tourism in places like Cape May and Long Branch as early as the late 1700s. Returns to investment on the Jersey Shore flourished during late nineteenth and early twentieth century, as resort towns like Atlantic City became important symbols for the increasingly prominent American Dream of upward mobility. Rising disposable incomes for the middle and working classes also increased the importance of second-home ownership, which was promoted by local government and private-sector boosters, particularly after World War II. Even as second-home developments proliferated and upgraded, older resorts declined to a point where large-scale redevelopment, often directed by local government, came to prominence. In this way, the Shore's history is one of progressive commodification of natural amenities, whereby land, water, and natural cycles are transformed into products that can be privatized, sold, and consumed. The damage to the region caused by Sandy, and the region's subsequent redevelopment, provide a unique opportunity for real estate and tourism developers to make a lot of money, even at the expense of the natural ecosystem and people with modest means.

Nature Tourism as the Economic Engine of the Jersey Shore

Tourism generated nearly $40 billion in 2012, accounting for roughly 7 percent of New Jersey's total economy, directly employing more than 300,000 of the state's residents and yielding $4.5 billion in state and local tax receipts.[1] By employment, tourism is New Jersey's fifth largest private economic sector (after health care, retail, professional services, and finance) and accounts for nearly one in ten jobs in the state.[2] The New Jersey Division of Tourism and Travel asserts that if the state were to forgo tourism revenues, each household in the state would need to pay $1,420 more in taxes each year.[3] Although tourism does take place in other regions, the four coastal counties (Atlantic, Cape May, Monmouth, and Ocean) account for more than 50 percent of direct sales in tourism statewide, more than 40 percent of direct employment in tourism, and almost 45 percent of state and local tax receipts from tourism (see table 3.1). Jersey Shore tourism,

TABLE 3.1.

Economic Impact of Shore Tourism, 2012

	Direct sales (millions of dollars)	*Direct employment*	*State/local tax receipts (millions of dollars)*
Atlantic County	7,594	62,435	835.5
Cape May County	5,241	24,464	481.1
Monmouth County	2,089	20,289	269.1
Ocean County	4,198	26,101	433.6
Shore total	19,122	133,289	2,019.3
New Jersey total	37,555	318,560	4,490.4
Shore as % of New Jersey	50.65	41.84	44.97

Tourism Economics, *The Economic Impact of Tourism in New Jersey: Tourism Satellite Account Calendar Year 2012* (Trenton: New Jersey Division of Travel and Tourism, 2012), *Source*: Adapted from http://www.visitnj.org/sites/default/master/files/2012-nj-tourism-ei-state-counties-v0701.ppt, 68, 71, 73.

in other words, is not just an emotional or recreational concern but one of the most important economic activities in the state. Superstorm Sandy sent shock waves through the state's economy.

Tourism is unlike other economic sectors in that it spans multiple types of businesses, the most important of which include lodging, food and beverages, retail trade, recreation, and transportation. Tourism-related businesses include both small mom-and-pop shops and enormous multinational corporations. Tourism also relies heavily on nonvisitor expenditures, including government support, capital investments, and spending by second-home owners. State support of the tourism industry, including the entire budget of the state's Division of Tourism and Travel and "other budget items in broad support of tourism," amounted to more than $116 million in 2011 and $123 million in 2012.[4] Capital investments, such as the construction of hotels, accounted for roughly $1.4 billion annually in New Jersey in both 2011 and 2012.[5] Purchases by "nonvisitors," especially second-home owners, were estimated to generate an additional $187 million in 2011 and $192 million in 2012.[6] Together with direct spending by visitors, tourism was a $39.5 billion industry in New Jersey in 2012.[7]

New Jersey tourists spend their money in lodging (30 percent), food and beverages (23 percent), retail (18 percent), transportation (18 percent), and recreation (11 percent),[8] although this varies by Shore county (see tables 3.2 and 3.3). Atlantic County's tourism industry is dominated by corporate casino hotels, which capture revenues associated with gambling, food, beverages, retail, and other recreation, all of which are officially reported as "lodging" since hotels fall into this sector. In contrast, Monmouth County's tourism revenues rely more heavily on food and beverage sales than on lodging, but revenues are actually

TABLE 3.2.

Direct Sales in Tourism-Oriented Sectors, 2012 (in millions of dollars)

	Lodging	*Food and beverages*	*Retail*	*Transportation*	*Recreation*	*Total*
Atlantic	4,429.5	1,244.5	959.0	585.7	375.7	7,594.4
Cape May	2,232.7	1,130.0	879.9	410.5	588.3	5,241.4
Monmouth	436.8	525.7	390.0	297.9	438.8	2,089.2
Ocean	1,334.1	994.8	811.3	504.6	553.2	4,198.0
New Jersey total	11,401.8	8,615.0	6,682.4	6,804.0	4,251.6	37,754.8

Source: Adapted from Tourism Economics, *The Economic Impact of Tourism in New Jersey*, 70.

TABLE 3.3.

Direct Sales in Tourism-Oriented Sectors as Percentage of County Tourism Direct Sales Totals, 2012

	Lodging	*Food and beverages*	*Retail*	*Transportation*	*Recreation*	*Total*
Atlantic	58.41	16.41	12.64	7.72	4.95	100.00
Cape May	42.60	21.56	16.79	7.83	11.22	100.00
Monmouth	20.91	25.16	18.67	14.26	21.00	100.00
Ocean	31.78	23.70	19.33	12.02	13.18	100.00
New Jersey total	30.20	22.82	17.70	18.02	11.26	100.00

Source: Adapted from Tourism Economics, *The Economic Impact of Tourism in New Jersey*, 70.

spread fairly evenly across all five sectors. These figures indicate that the tourist economy varies in each county, with Atlantic and Cape May Counties more reliant on overnight visitors than on day-trippers and owners of second homes.

This pattern of tourism is reflected in employment patterns across Shore counties. Employment in Atlantic County is most concentrated in the tourism industry. Cape May County is a close second in this respect, while Monmouth and Ocean demonstrate much broader employment options (table 3.4). Shore tourism is highly seasonal, so employment in this sector is typically less stable than in other industries. Unemployment rates underscore the precariousness of specialized labor markets in Atlantic and Cape May Counties, although it should be noted that because of the relatively small size of the labor force in the Shore's two southern counties, there are actually more unemployed individuals in both Monmouth and Ocean Counties than in the two southern counties combined. Given their greater dependence on tourism, the smaller, southern Shore counties are particularly vulnerable to weather-related disruptions to the industry, while the northern Shore counties are more economically resilient. In this sense, New Jersey's tourism industry was fortunate that Sandy spared much of Atlantic and Cape May Counties.

Sandy struck after the 2012 summer season, so its impact on tourism receipts was relatively small. The greatest immediate economic impact of the storm involved the closure of Atlantic City's twelve casinos for one week. In the previous year, the New Jersey Casino Control Commission reported that Atlantic City casinos earned $2.95 billion in taxable revenues, generating more than $236 million

TABLE 3.4.

Employment Statistics for Shore Counties, 2012

	Employment	*Tourism employment*	*Tourism as % of total employment*	*Unemployment rate %*
Atlantic	117,700	62,435	53.05	13.5
Cape May	50,400	24,464	48.54	13.4
Monmouth	304,900	20,289	6.65	8.9
Ocean	244,100	26,101	10.69	10.3
Shore total	717,100	133,289	18.59	10.49
New Jersey total	4,159,300	318,560	7.66	9.5

Source: New Jersey Department of Labor and Workforce Development, Office of Research and Information, "Table 6: New Jersey Resident Population by Municipality, 1930–1990" (Trenton: State of New Jersey, 1990), http://lwd.dol.state.nj.us/labor/lpa/census/1990/poptrd6.htm; and Tourism Economics, *The Economic Impact of Tourism in New Jersey*, 71.

in tax revenues.[9] The Division of Tourism and Travel reported that Sandy-related closures and the storm's aftermath led to a 28 percent drop in revenues from the previous year for the month of November 2012.[10] The evacuation of entire communities on the barrier islands, the displacement of hundreds of thousands of others as a result of either storm damage or power outages statewide, and the influx of disaster workers and volunteers translated into an 80 percent increase in bed taxes in November and December 2012.[11] These short-term blips did not compare to what the consulting group, hired by the New Jersey Division of Tourism and Travel to analyze the economic impact of tourism, called "The $22 Billion Question":[12] Would the Jersey Shore would be "ready" to accommodate the more than thirty million visitors expected in the summer of 2013?[13]

Although Sandy caused damage up and down the Jersey Shore, the most severe damage to tourist facilities was in Monmouth and Ocean Counties, especially Ocean County. Although these two counties are less focused on tourism than are Atlantic and Cape May Counties, tourism still directly provides more than forty-six thousand jobs in Monmouth and Ocean, as well as seven hundred million dollars in annual tax revenues. Given these numbers, there is little question that the Shore would need to dedicate recovery efforts on the income-generating elements of the Shore community. While acknowledging the devastating impact on individuals whose homes (and second homes) were damaged and destroyed, the Jersey Shore—and New Jersey as a whole—simply could not afford to delay the redevelopment of tourism facilities. For the same reason, there was little debate about whether tourism facilities should be rebuilt. Billions of dollars in federal assistance streamed into the region, and most tourist facilities, including all public boardwalks, were ready to receive visitors by the 2013 summer season. On May 24, 2013, in Seaside Heights, Governor Chris Christie officially opened the summer season by cutting a 5.5-mile ribbon that stretched along the Shore; it read "New Jersey: Stronger Than the Storm." He told *Today* show host Matt Lauer that "for probably 80 percent of the Jersey Shore, you won't notice the difference at all from last summer."[14]

Having visited many of the public boardwalks and amusement centers from Ocean City to Keansburg throughout the 2013 season, I can confirm that much of the tourist Shore was, in fact, ready for business and actively receiving visitors. Boardwalks in Atlantic and Cape May Counties, such as those in Ocean City and Wildwood, barely showed any signs of storm damage. Point Pleasant's rides were open for the spring preseason, and repairs to the boardwalk were complete by May. Repairs and upgrades to the Belmar and Asbury Park boardwalks were so complete as to be barely noticeable, although the absence of permanent buildings on the east side of the Belmar boardwalk counted among the 20 percent "difference" from the previous season. A few rides and the Haunted House in at the Keansburg Amusement Park remained closed for the 2013 summer season, but even this hard-hit Bayshore destination was mostly operational.

3.1. Boardwalk, Seaside Heights, summer 2013. Photo by author.

3.2. Claw game with "Restore the Shore" prizes in Seaside Heights, summer 2013. Photo by author.

The most notable exception to the restoration of tourism facilities on the Shore involved the twin amusement piers at the boardwalk spanning Seaside Heights and Seaside Park. The Casino Pier in Seaside Heights collapsed during the storm, leaving the Star Jet roller coaster and other rides in the surf. Just to the south, the Funtown Pier in Seaside Park also suffered serious damage. Some businesses in Seaside Heights had opened before the end of 2012, catering to people involved in the recovery effort and tourists interested in seeing storm damage firsthand. The boardwalk itself was replaced in its entirety by the start of the summer season, and most of the storm debris had been removed. Both piers remained closed, but they became symbols of the recovery effort and earned a great deal of media attention, including live national television coverage of the sinking of pilings for the new Casino Pier in February 2013 and a full-color photo on the front of the *New York Times* travel section on August 2, 2103.[15] The euphoria ended for the Funtown Pier, however, during the "Local's Summer" of September 2013, when a fire—probably caused by degradation of electrical wires due to exposure to sand and salt water during Sandy—spread rapidly through the pier and adjacent businesses.[16] The remnants of the pier and eight blocks of businesses were affected, with more than twenty structures reduced to ashes.

The Growth Machine

Clearly, the Shore is economically dependent on tourism. This is both promoted and reinforced by multiple interests, including resort owners, small business owners, real estate interests, local government, and homeowners, among others. The coalition of economic interests varies historically and geographically, but tourism-related economic activities have expanded over time to include nearly all of the 127-mile coastline. Initially, eighteenth- and nineteenth-century resorts in Cape May and Long Branch were developed for elites. At the turn of the last century, resorts like Atlantic City and Asbury Park served to draw large numbers of the newly affluent white ethnic middle classes of the New York and Philadelphia metropolitan regions. Relatively new to their class status, these visitors desired a public stage on which to demonstrate their affluence, and the boardwalks of the Jersey Shore served this function perfectly. As technology and the middle classes became more affluent and less insecure about their social standing, second-home development intensified along the Jersey Shore, particularly in waterfront communities surrounding historic resort centers and along the region's bays, rivers, and inlets. These second-home owners (and visitors who rented private homes during the summer season) were attracted to newer, more family-friendly tourist facilities that were more easily surveilled and controlled than older resorts were. This led to the decline of older resorts, most notably Atlantic City and Asbury Park, whose aging facilities, poor, nonwhite locals, and declining public safety contrasted with middle-class recreational ideals. The derelict condition of these resorts and growing regional affluence in the late twentieth century has led to opportunities for strategies of economic recovery, such as state-led redevelopment (for example, Atlantic City), public-private partnerships (for example, Long Branch), and gentrification (for example, Asbury Park). Sandy hit a Shore that was already in transition toward greater affluence, and many predict that the storm will concentrate and consolidate those processes, particularly in locations with high or potentially high real estate values.

Although few consider the Jersey Shore to be a "nature tourism" destination, University of Oregon sociologist John Bellamy Foster has written extensively about how commodification of nature involves converting a naturally occurring product into something that is traded in a competitive market. Tourism is different from other economic sectors only in that there is not a physical product that is traded; rather, an experience is exchanged in a marketplace. Moreover, the coastal tourism product cannot be separated from exposure to environmental hazards, as a warm ocean such as that off the Jersey Shore is a natural incubator for storms. Coastal tourism, both at the Jersey Shore and elsewhere, must accept the cost of storm damage as "normal," or intrinsic to the system. At the same time, it also makes economically rational sense for those who generate private profit from tourism to socialize the risk and cost of storm damage. As such, political and economic structures of this region are organized around the promotion of tourism, and the preparation for and response to storm damage is part of a public regional commitment. Thus, while it may make no environmental sense to rebuild structures destroyed in storms, there are powerful interests invested in Shore tourism, and these share widespread political and cultural support. This helps explain why there has been such a strong emphasis on rebuilding the boardwalks in the wake of Sandy; it is not just feel-good stuff, it is big business.

In the late 1960s, sociologist Harvey Molotch analyzed the effects of another disaster on a populated coastline. Along with other residents of the Santa Barbara, California, area, Molotch watched with horror as an oil spill despoiled their beautiful Pacific coast. Molotch observed that local and state government officials and local economic actors joined forces to promote economic development but generally failed to protect the natural environment.[17] Molotch called this coalition a "growth machine" and described it less as a conspiracy of elites than a rational and collaborative process within capitalist economies. Local businesses understand that they benefit economically by cooperating with other businesses in the same geographical region, forming organizations like chambers of commerce that promote regional businesses directly through joint advertising and some facilities (such as in regional malls), and indirectly through political influence. Political influence is critical for the infrastructural developments necessary for the continued successes of private businesses; utilities, roads, parking, and many other public services affect how successful individual businesses are. However, individual business owners, and even larger corporate actors, will not or cannot shoulder the cost of this infrastructure, often because it is simply too expensive, especially for small mom-and-pop businesses, but also because it is *possible* to shift the costs of such infrastructure to the public, thus spreading the cost among the general population, not just business owners. Molotch refers to these business leaders as place entrepreneurs: those who have vested interests in promoting economic growth in a specific geographic location.

Politicians, who are also tied to geographic place and are often drawn directly from the local economic elites, generate tax revenues by promoting the growth of the private sector and the proliferation of new residences, particularly high-value residences. Political constituents have historically supported the creation of new jobs, new "ratable" (that is, taxable) businesses, and new additional property tax receipts—provided that new residents can be expected to contribute more to the existing place than they will require in public services. The development and growth of these new income streams allow municipalities, counties, and states to increase public services, including recreational amenities, improved public schools, professional police and fire services, better roads, and a variety of other locally desirable benefits. The local and state political apparatus need not be co-opted or corrupt to focus its efforts on promoting economic and real estate development within its political boundaries; on the contrary, the growth machine logic underscores how private economic interests typically partner with governments in a way that provides at least some benefits to the private sector, politicians, and the general public. Consistent with capitalist organization of the economy, the revenue generated in this process flows unevenly, with private businesses capturing the lion's share, often legitimated with the controversial assumption that this money will eventually trickle down through the local economy through taxes, jobs, and local purchases.

In addition, place entrepreneurs have rational reasons for promoting civic pride and boosterism, as long as these support continued economic growth. Civic pride, or in Molotch's term, place patriotism, allows one place to distinguish itself from competing places to attract private and public investments. Place patriotism is particularly important when the economy is not booming, because economic competition and constituent demands are both greater. The creation of a proud, engaged, and well-functioning community becomes a competitive advantage. Place entrepreneurs, both in the private sector and in the local government, thus have an economic motive to encourage and fund activities and institutions that foster emotional attachment to place. Civic clubs, youth recreation, museums, festivals, parades, and other cultural institutions and events are strategic investments for place entrepreneurs. These bring community members into the growth machine coalition, because residents tend to experience their homes and home towns as more than simply real estate; in Marxist terminology, they respond to the "use value" of place rather than just its "exchange value." Residents are particularly receptive to activities that foster social integration and community building, even without an economic incentive to do so, although civic engagement and regional economic growth may also increase property values.

Despite these positive outcomes, Molotch found that growth machine alliances have concentrated political and economic power and done a poor job of protecting noneconomic interests, including the environment. The logic of the growth machine favors the interests of private business over the public good and may lead local governments to favor business interests over individual residents or the public good. If environmental protection is considered a public good, protection may be contrary to the interests of the growth machine. Land preservation, for example, literally reduces the space available for economic development. Increased regulation of environmental emissions, in another example, suggests that noneconomic interests (such as public health) take primacy over capital accumulation. Environmental protection is generally antithetical to the logic of the growth machine alliance, and only with political opposition to the growth machine is environmental protection possible. Elite residential communities, such as Santa Cruz, California, and Aspen, Colorado, as well as affluent suburbs in New Jersey, have successfully challenged growth machines in their local political institutions. However, these successful anti-growth movements draw from wealth generated *outside* the local economy. It follows, then, that the more dependent residents are on the local economy, the more receptive they will be to growth machine politics and thus less receptive to environmental protection.

When their economies are tied to the environment, as is the case in nature tourism destinations, the growth machine can sometimes run counter to the long-term economic health of the place which it seeks to develop. Rutgers sociologist Thomas K. Rudel found this to be the case in the Ecuadorian Amazon, where a coalition of state actors, foreign aid agencies, and small farmers worked to carve agricultural colonies from an expansive rainforest.[18] Although the colonies initially boomed and colonists made small fortunes, removal of the forest created ecological degradation that eventually undermined the agricultural economy as well as depleted the forest itself. Within thirty years, most colonist households pursued economic strategies other than farming, including migration to national cities and abroad.[19] In the same way, growth machines (or coalitions) have created cycles of booms and busts throughout U.S. history, and are responsible for a variety of environmental problems, including the endangerment, local extinction, and absolute extinction of commercially valuable animals, the depletion of mineral and forest resources, chemical contamination, and large-scale soil erosion. Political intervention can stem the boom-and-bust cycle, but historically, the political will to put the brakes on growth only materializes when depletion or degradation of the resource threatens the economic viability of the community itself.[20]

Tourism-dependent economies share many characteristics of resource-dependent communities, particularly when what attracts tourists is a natural feature, such as ocean beaches. Place entrepreneurs lobby to develop natural features through private investment (especially in real estate), political coalitions, infrastructural development, advertising, and civic institutions and activities. Tourism destinations must compete with other destinations to attract visitors, and this process has historically created a boom-and-bust pattern known as the "resort cycle" whereby the promotion of destinations leads to increased tourism, which degrades the nature of the environmental attraction and typically makes a destination less exclusive.[21] Both the changed natural attraction and the larger number of less elite visitors lead to a degradation of the "natural" experience, leading to a decline in the desirability of the location and thus economic activity. Like resource-dependent communities, an unregulated growth machine is likely to undermine the environmental base of nature tourism.

Environmental sociologists also like to point out that the cost of environmental degradation should be internalized to the cost of production, rather than treated as external to the economic equation. The process of failing to account for environmental externalities that has encouraged the degradation of ecosystems through pollution and depletion.[22] Thus, the true cost of any given product includes resource inputs, labor, capital investments, energy, waste disposal, and mitigation of ecological damage and threats to human health caused by any part of production. When these true environmental costs are taken into account, the actual costs of production are often much, much higher than current consumers pay. Although the processes of economically evaluating environmental services and estimating the cost of reduced ecological functioning and human health are methodologically tricky, the main barriers to adding these costs ("internalizing them") are political. That is to say, people who have the power to continue to exclude these costs from the economic equation and write the rules of how economic costs and benefits are calculated allow for these costs to be ignored entirely or are able to transfer the costs to others, usually the public and future generations.

Most research on the externalization of environmental costs has focused on the production of physical commodities like agricultural products and petroleum, but there is no reason the same basic principles cannot be applied to the capitalist production of services, including those associated with tourism on the Jersey Shore. As an example, to provide amusement rides for children on one of New Jersey's boardwalks, a business must provide some basics: (1) a physical location, which requires either the purchase or renting of land; (2) the actual rides, as well as the gates, ticket booths, and other fixed capital costs and maintenance; (3) a labor force capable of safely operating the rides and trustworthy enough to manage money received from customers; (4) energy to run the rides and light up the park; and (5) incidental items, such as the actual tickets themselves. Typically, business owners also pay for advertising costs, which allows them to inform and persuade potential customers to visit their amusements; they need to do this because entertainment is a discretionary item that may not be purchased at all, and because of the competitive nature of the amusement-destination marketplace, where one park must lure potential customers away from others. Groups of small-business owners who occupy a concentrated physical space may pool resources to advertise, thus reducing the burden of advertising costs, a form of cooperative capitalism consistent with the growth machine. Modern business owners also pay, although frequently politically resist, regulatory costs, such as paying employee benefits (including those mediated through government programs such as social security), insurance costs, and regulatory compliance costs (such as providing public restrooms, safe working environments, and waste disposal). An amusement park owner must bring in enough money from visitors to offset all of the costs noted here and to provide surplus profit that compensates for the owner's time, investment, and risk. In 2012, a team from Rider University found that failure rate of small New Jersey businesses (excluding home-based businesses) could be as high as 47.7 percent by the second year,[23] suggesting that starting and running a business is indeed a very risky proposition.

Given this situation, it is not surprising to find that business owners try to offset some of these costs. The social contract that arose in the wake of the Great Depression obligated the nation as a whole to provide for the basic infrastructure of the private economy. For example, the federal and state governments channel tax dollars into creating roads, bridges, trains, and airports that serve both personal and business interests. Public infrastructure costs in seaside communities are exceptionally high, however. Subsidized public investments include beach stabilization and replenishment efforts, dune construction and management, a transportation structure that requires long bridges, deep pilings, extensive soil stabilization, and (to date) highly subsidized hazard insurance. Coastal areas also benefit from unique public recreational amenities such as boardwalks, piers, and monitored beaches.

New Jersey is somewhat unusual in the United States in that it attempts to recoup at least the cost of beach maintenance through user fees, rather than spreading the costs among users and nonusers alike. User fees take the form of beach badges, which require people to pay to use a public beach either daily, weekly, or for an entire season. By state law, the cost of beach badges cannot be higher than the cost of beach maintenance and operation, and badge fees cannot be used to subsidize a municipality's general budget. Despite these provisions, badge fees are highly controversial in New Jersey because they add a direct cost that is not present in most other states' beaches and therefore make New Jersey beaches more expensive to consumers relative to other states. Fees may also have a regressive effect on consumers, whereby the poorest members of the population will avoid the beach because of the added cost.

To fully internalize the cost of maintaining beaches in New Jersey by user fees alone would make the cost of beach badges far higher than most users could possibly afford. Table 3.5 indicates the revenues generated through beach badges in Monmouth and Ocean Counties, which collectively come nowhere near the $102 million that the U.S. Army Corps of Engineers has allotted for restoration and stabilization projects in Monmouth County alone in the wake of Superstorm Sandy. Based on the cost of beach badges and assuming a 104-day beach season for ten years (one week each of May and September and

TABLE 3.5.

Public Badge Fee for Monmouth and Ocean Counties, Before and After Sandy, from North to South

	Sales 2012 (in dollars)	*Sales 2013 (in dollars)*	*% Change*
MONMOUTH COUNTY			
Sea Bright	283,257	189,420	-33
Monmouth Beach	—	—	—
Long Branch	1,368,497	1,232,663	-10
Deal	61,298	29,994	-51
Allenhurst	1,303,970	1,263,551	-3
Loch Arbor	200,264	—	—
Asbury Park	606,554	713,483	+18
Ocean Grove	—	—	—
Bradley Beach	1,268,900	1,287,502	+.05
Avon	1,125,942	962,317	-15
Belmar	2,230,585	2,001,995	-10
Spring Lake	1,364,916	1,427,960	+5
Sea Girt	879,588	893,947	+2
Manasquan	1,408,506	1,055,338	-25
OCEAN COUNTY			
Point Pleasant Beach	—	—	—
Bay Head	—	—	—
Mantoloking	—	—	—
Brick	457,728	190,230	-58
Lavalette	709,790	572,183	-19
Toms River	782,673	125,000	-84
Seaside Heights	1,600,464	1,156,714	-28
Seaside Park	1,558,905	1,326,780	-15
Berkeley	74,199	68,623	-8
OCEAN COUNTY (LBI)			
Barnegat Light	202,675	203,499	+.4
Harvey Cedars	215,451	210,205	-2

(*Continued*)

(*Continued*)

	Sales 2012 (in dollars)	*Sales 2013 (in dollars)*	*% Change*
Surf City	549,292	488,309	-11
Ship Bottom	730,935	566,150	-23
Long Beach	1,627,340	1,399,430	-14
Beach Haven	462,255	410,000	-11

Source: Adapted from *Asbury Park Press*, "Graphic: Were Beach Badge Sales Up or Down This Year in Your Town?" *Asbury Park Press*, August 30, 2013, http://www.app.com/interactive/article/20130830/NJNEWS/130830003/Graphic-Were-beach-badge-sales-up-down-year-your-town-.

Note: Ocean Grove in Monmouth County and Point Pleasant Beach and Bay Head in Ocean County did not provide data to the *Asbury Park Press* because beaches are operated by a private enterprise. Data from Monmouth Beach and Loch Arbor in Monmouth County and Mantoloking in Ocean County was not available at the time of printing in the *Asbury Park Press*.

the entire months of June, July, and August), *each* of the thirteen Monmouth County beach towns would need to sell an average of 943 *daily* beach badges *every day* for ten years to recover just the cost of the Corps of Engineers projects. This leaves no money for lifeguards or beach maintenance, which may not break even, even in years without massive storm-related damage. For example, Long Branch estimated that it earned $1.6 million in badge sales in 2011, but this left the community with a net deficit of $300,000 for beach operations and maintenance.[24] Considering that many beachgoers pay considerably less than the daily rate by going to the beach on weekdays, by buying seasonal or weekly badges, or by being seniors or children, and that some summer days are rainy or cold, even in a "normal" year, it is unrealistic to expect user fees to cover operational expenses and to have beaches available to the public at an affordable cost.

Transportation requires another large public investment that benefits individual as well as business interests along the Shore. The reconstruction of State Highway 35 from Point Pleasant Beach to Island Beach State Park is expected to cost more than $200 million.[25] In the recovery period after Sandy, the New Jersey Department of Transportation released plans to spend more than $900 million on road and bridge improvement projects in Monmouth and Ocean Counties between 2012 and 2021, with more than 70 percent ($666.591 million) allotted to the less-populated Ocean County.[26] These figures amount to an investment of

$382.13 for every resident of Monmouth County and $1,156.14 for every resident of Ocean County. Although these figures are not out of proportion for expenditures across the state, they underscore how the public sector subsidizes the infrastructure required by the tourism industry.

Like all other sectors of the economy, tourism businesses on the Jersey Shore do not incorporate the costs of associated environmental services, environmental degradation, and health risks. Significantly, seaside tourism businesses are highly dependent on inputs that come either "free" from the natural environment or highly subsidized from public funds. For example, "free" services from the environment include an aesthetically pleasing landscape (beach and ocean), naturally cool breezes and fresh air, and environmental amenities that will attract potential customers independent of advertising (for swimming, sunbathing, surfing, or fishing). Many of these environmental services are partially regulated by state agencies, such as the New Jersey Department of Environmental Protection (DEP) or the U.S. Environmental Protection Agency (EPA), and are thus partially paid through taxes, fees, or fines. However, persistent and worsening problems, such as those associated with climate change, marine and terrestrial biodiversity loss, and both air and water pollution, indicate that existing regulation has not curbed net degradation in environmental systems. Thus, even if regulatory structures attach some costs to these "free" environmental services, the true cost of maintaining environmental services is not yet balanced by these obligations.

The Resort Cycle on the Jersey Shore

How did the Shore come to be so economically dependent on a vulnerable sector like oceanfront tourism? Growth machines exist up and down the Jersey Shore, but the particular nature of coalitions and the type of tourist facilities have varied across time and space. There are two economic boom periods for the Jersey Shore resorts: the halcyon period of resorts that were developed by the end of the nineteenth century or very early in the twentieth century, and the redevelopment period of renewed reinvestment in Shore resort communities at the end of the twentieth and beginning of the twenty-first centuries. These booms are separated by the bust of the great Shore resorts like Asbury Park and Atlantic City, as white middle-class consumers pursued recreation elsewhere. In both boom periods, place entrepreneurs capitalized on aspects of the tourist economy that provided that greatest opportunity to generate wealth from the Shore's natural amenities, but they also sought to reduce economic risk through public investment, particularly in infrastructure but also through storm-damage mitigation and repairs.

The Halcyon Period

The Jersey Shore's earliest resorts were very different from modern tourist sites in a variety of ways. First, early resorts were remote and relatively difficult to access. A map of railroads in 1870 shows rail connections to relatively few locations on the Shore.[27] Because of its industrial history, the Raritan Bay shore was most heavily serviced by railroads during this time period. The Camden & Amboy terminus at South Amboy was an important link between New York and Philadelphia, but it did not help develop seaside resorts. Likewise, the New Jersey Southern Railroad provided branch line service to Port Monmouth, Red Bank, and Shrewsbury; to Atlantic Highlands via Long Branch; and to Toms River. Among these communities, only Long Branch had a reputation as a seaside resort. However, there were additional rail lines to Sea Girt (the Freehold & Jamesburg line), Atlantic City (Camden & Atlantic), and Cape May (via the West Jersey Railroad and the Camp May & Millville).

According to nineteenth-century Monmouth County historian Franklin Ellis, in 1790 or 1791 Lewis McKnight purchased a homestead in Long Branch with the intention of developing a seaside resort, and by 1792, hotel proprietors were advertising Long Branch resorts in Philadelphia.[28] A visitor in 1840 recalled an active seashore resort, including hotels, bathing facilities, beach bowling, and "colored cooks from the South."[29] Ellis reported that by 1860 Long Branch hotels could accommodate more than four thousand guests, who avidly swam in the ocean and enjoyed the offshore breezes from the pine forests in the west.[30] As they do now, resorts faced special problems as a result of being located next to the ocean on the rapidly eroding headlands: the site of the original Monmouth House hotel, built in 1848, was already a half mile out to sea in 1885.[31]

At the far southern tip of New Jersey, English whalers had settled a village on Cape Island in 1630s, but Cape May claims to be the "nation's oldest seashore resort," with resort facilities in full swing by 1766.[32] Cape May resorts drew initially from Philadelphia and Baltimore but expanded to include regular visitors from the U.S. South and New York City. In 1878, at the time of a devastating fire that leveled almost the entire hotel area of the island, Cape May decided not to compete with growing competition from larger, more garish seaside resorts such as Atlantic City, and instead rebuilt in the smaller-scale Victorian style for which it is still recognized and for which the entire city of Cape May is listed on the National Registry of Historic Places.

By 1887, Jersey Shore tourism was booming and rail access to the region had changed dramatically.[33] Rail lines now served the entire Monmouth County coast and continued down the Barnegat Peninsula to Seaside Park, where one crossed the Barnegat Bay near the location of today's Toms River Bridge (State Highway 37). Railroads also traversed Long Beach Island from the Barnegat Inlet to Beach Haven and Absecon Island from Atlantic City to Longport. In Cape May

County, railroads served all barrier island communities from Ocean City to Sea Isle City, and there was a branch line to the Wildwoods and several stations in Cape May. The railroads in many ways engineered the early development of seaside resorts, both directly, as in Atlantic City, where the railroad was among the largest landowners at the resort's formation, and indirectly, as in Monmouth County, where railroads facilitated travel for urban beachgoers.

Railroad developers identified Absecon Island in the mid-nineteenth century as an ideal place for a seaside resort, given its proximity to Philadelphia. Although the first facilities were developed before the arrival of the Camden & Amboy rail service in 1854, Atlantic City became the "Queen of Resorts" as a direct result of increased rail service and associated real estate development. Like Cape May and Long Branch before it, Atlantic City's main attraction was its broad beach, which could alleviate the crushing heat and humidity of northeastern summers. The town was designed from the outset to accommodate visitors who arrived by train, and in fact it was accessible only by train until 1870. As a consequence, Atlantic City was not built to accommodate horse-drawn transportation, but instead developed as a truly walkable city. Not coincidentally, it was here that the boardwalk was constructed as a means of allowing bathers easier access to, from, and along the beach. The boardwalk rapidly grew as a linear focus for shops and attractions along the beachfront, then grew into a destination where visitors could promenade. Atlantic City's popularity is most closely associated with the turn of the twentieth century through the era of Prohibition, when the boardwalk attracted literally millions of visitors to view the unique hotel architecture, consume both wholesome and exotic entertainment, and walk amid movie stars and celebrity gangsters.

As the resort became increasingly accessible across class, ethnic, and racial lines, however, its attractiveness to elites began to wane. By the middle of the twentieth century, it was well into decline. Reconstruction of boardwalk facilities from the Great Atlantic Hurricane of 1944 and the Ash Wednesday Storm of 1962 did not attempt to fully replace what had been lost. Despite the best efforts of Atlantic City's growth coalition, the city was destitute and partially abandoned when state voters approved casino gambling as a last-ditch effort to revive tourism in 1976. When Resorts International opened Atlantic City's first casino just two years later, the state and city governments worked alongside private investors as the prime movers of the city's redevelopment. Net gains from casino gambling in Atlantic City remain unclear and debatable,[34] but the activist role of the local government in promoting redevelopment of troubled Shore resorts characterized growth coalitions in this period.

Monmouth County beach towns had also developed rapidly in the middle of the nineteenth century. The earliest resort in the county was undoubtedly Long Branch, which transformed over the century from an elite seaside health resort attracting seven presidents (thus the name of the eponymous state beach) to a destination more associated with gambling at the Monmouth Racetrack.[35] After gambling was banned in New Jersey in 1898, in part due to the influence of Asbury Park developer-turned-senator William Bradley, Long Branch reoriented its tourism industry toward more wholesome entertainment. This centered on a long steamboat and fishing pier that was converted to an amusement pier in the early twentieth century. The beach and pier attracted more working-class visitors, who also patronized dance halls and illegal gambling venues throughout Long Branch, which by 1921 was described as a "veritable Monte Carlo."[36] With the influx of military personnel prior to and during World War II, Long Branch increasing offered more adult-oriented alternative attractions. The Long Branch Pier and boardwalk had rides, arcades, pools, beach clubs, and other similar amenities, but it was better known for its bars and dance halls. Its reputation for attractions like this, along with the increasingly working-class, white ethnic, and African American population, made Long Branch less attractive to affluent tourists. When the Great Atlantic Hurricane of September 1944 destroyed "every single structure on Long Branch's beachfront,"[37] many of the affected businesses were already in decline and did not rebuild.

In contrast to Long Branch, Ocean Grove was founded by Methodist minister William B. Osborne in 1870; he sought to build a seaside religious "camp" that would allow New York City's pious teetotalers a retreat from the disorder of city life. The planned community that developed allowed worshipers of varied economic means access to the camp. People with modest incomes could purchase small plots, including seasonal tent homes. More affluent visitors could purchase summer homes built using staggered setbacks from the property lines to allow more homes to have direct views of the ocean. The community strictly regulated vehicular traffic on Sundays and prohibited the sale of alcohol, both of which traditions continued into more recent times.[38] The sedate beach town offered religious retreat, but one entrepreneurial worshipper discovered economic opportunity as well.

An early investor in Ocean Grove, William Bradley, wandered just past a lake that demarcated the community's boundary in 1871 and decided to build an adjacent, but secular resort, which he named Asbury Park after the first Methodist bishop in the United States. The property's owner agreed to sell Bradley five hundred undeveloped acres, allowing Bradley to develop an entire city from scratch. Bradley envisioned a family-oriented seaside resort, with hotels, boardwalks, and amusements, connected to population centers in the north by rail. By 1888, Bradley had attracted Ernest S. Schnitzler, who had previously owned and operated a merry-go-round in Atlantic City, to build what would become Palace Amusements just off the boardwalk. Palace Amusements began as a merry-go-round but eventually included an observation tower, Ferris wheel, crystal maze, dark ride ("tunnel of love"), swimming pool, haunted house, arcade, funhouse, and beach facilities.[39] By 1902, Palace Amusements was described as "the largest, most unique and most complete [amusement center] under one roof of all found on the Atlantic coast."[40] Schnitzler sold Palace Amusements in the 1920s to August M. Williams, who expanded amusements to the beachfront Casino. Most Asbury Park facilities were located on the inland side of the boardwalk, so although the Great Atlantic Hurricane of 1944 severely damaged the Asbury boardwalk, the Casino and other landmark buildings, including Palace Amusements, survived.

Palace Amusements continued to expand into the 1950s, when then-owners Edward Lange and Zimel Resnick commissioned Leslie W. Thomas to paint exterior murals to attract tourists. The most famous of these was modeled after a clown face originally used by Coney Island amusement entrepreneur George Cornelius Tilyou; the clown face became known as Tillie and has served since as a symbol for Asbury Park.[41] Palace Amusements, however, has not survived. Asbury Park, like many older resorts, became less attractive after people stopped traveling to the Shore on trains. The smaller boardinghouses that had been the pride of the halcyon period did not offer enough space, parking, or privacy for consumers increasingly accustomed to suburban amenities. In Asbury Park, residents willing to occupy the small rooms with common facilities were increasingly drawn from socially marginal groups: the deinstitutionalized, the mentally ill, and the addicted.[42] This made the town even less attractive for tourists, leading to a spiral of disinvestment and abandonment. In 1988, Palace Amusements closed its doors on a boardwalk largely devoid of businesses or tourists.

The Redevelopment Period

Asbury Park was not the only resort in economic straits by the latter decades of the twentieth century; in fact, most of the early resorts struggled. Historian Bryant Simon offers three main reasons for the decline of Atlantic City, which apply to all of the early resort towns: aging facilities, poor race relations between whites and growing minority populations, and changing consumer preferences. Local business owners and politicians pursued a variety of strategies to attract new visitors and promote redevelopment throughout the 1970s and 1980s, with varying degrees of success. These strategies have shaped the contemporary Shore, particularly in having attracted more affluent second-home owners to Monmouth County, and more mixed success with state-controlled gambling in Atlantic City. Redevelopment has also displaced poor, minority, and middle-income residents from more desirable locations along the Shore. In some places, like Atlantic City, local officials pursued a deliberate policy of removing or isolating poor neighborhoods. In other places, the private sector sent the poor on their way with equal efficacy, priming the Shore for more affluent residents. Sandy is likely to hasten the upscaling of the Shore, especially in Monmouth and Ocean Counties, where pressures for redevelopment and gentrification are strongest.

In the redevelopment period, there have been three important strategies for place entrepreneurs: (1) public-private partnerships, which leverage public funds to attract private investors; (2) gentrification, which promotes the replacement of poor residents with more affluent ones and is seen clearly in Ocean Grove and Asbury Park; and (3) second-home ownership, which has created new demands in the real estate market. These strategies for economic growth shape the way that the Shore's tourism "product" is offered to consumers, but they also increased the value of property and the number of people who may be affected by major storms like Sandy.

PUBLIC-PRIVATE PARTNERSHIPS. The Long Branch city government actively pursued redevelopment along its waterfront once its aging amusement pier and many neighboring businesses burned to the ground in 1987.[43] In its 1988 Master Plan, the City of Long Branch designated its oceanfront areas as the most important part of the city's redevelopment plan but made deliberate decisions to redevelop in zones rather than parcel by parcel. The city approved a Waterfront Redevelopment Plan in 1996, which was nationally recognized for incorporating state and regional zoning regulation, including the Coastal Area Facilities Review Act (CAFRA), which regulates new construction along the Jersey Shore (see chapter 5). Storm mitigation was built into the redevelopment plan; new buildings were to be built behind the natural bluff that characterizes the Monmouth headlands, which are reinforced by a bulkhead and beach groins in Long Branch. Even though sections of the boardwalk promenade did collapse in Sandy, redeveloped areas on the Long Branch waterfront suffered relatively minor damage.

The centerpiece of the Waterfront Redevelopment Plan was the creation of large-scale, mixed-use development in the area of the former pier. To create this, the city began to acquire properties, including the remnants of the pier and surrounding businesses, by having them legally declared as "blighted." Blight was vaguely defined by New Jersey's Local Redevelopment and Housing Law but had been even more broadly interpreted in case law to designate nothing more specific than an area in need of redevelopment. Thus designated, the city was able to procure the land around the pier through its right to eminent domain. Eminent domain involves the acquisition of private property by public agents for projects designed to provide social goods. Originally part of the legal process for acquiring land needed for roads, schools, and other such projects, eminent domain has become an important flashpoint in the conflict over the accumulation of private real estate capital on the Jersey Shore.

In April 2002, a private developer (Applied Housing of Hoboken, New Jersey), along with city and state officials, including Governor Jim McGreevey, broke ground on what would become Pier Village, which currently markets itself as "New Jersey's most luxurious oceanfront shopping and dining destination."[44] Along with upscale rentals and condominiums, Pier Village offers a recreational and commercial destination in what had been the dilapidated remnants of the former pier amusement center. Understandably, other developers were interested in reproducing this success.

In the Beachfront North zone of Long Branch, another private developer sought to purchase thirty-six small single-family homes, known locally as "bungalows," in order to create another large oceanfront complex. The majority of homeowners agreed to sell to the developer, but about one-third of owners declined. As it had with the dilapidated pier area, the city moved to designate these properties as "blighted" in 2003, in order to allow the redevelopment plans. Homeowners sued to prevent the condemnation, noting that many of their properties, while small, were well maintained and well utilized. Homeowners also questioned whether private redevelopment really constitutes a "public good" worthy of eminent domain.[45]

The Beachfront North homeowners were vindicated nearly five years after the attempted seizure when the New Jersey Supreme Court ruled that the city had not affirmatively demonstrated that the area was "blighted."[46] One year later, in 2009, the city announced that it had settled with remaining homeowners and would not pursue the condemnations.[47] When Sandy struck, the large multi-unit project redevelopment of Long Branch's waterfront stopped at the edge of the contested Beachfront North neighborhood. Sadly, the remaining bungalows are now dwarfed by huge new single-family homes and townhouses constructed after the individual properties sold earlier in the process were razed; ironically, or sadly, this is the exact type of aesthetically jarring redevelopment that the City of Long Branch had hoped to avoid by zone-oriented development. Boxed in by towering neighbors, many of the homeowners who fought eminent domain have since sold or are seeking to sell their bungalows (See photo 3.3.) Thus, while the "victory" of the homeowners over eminent domain in Long Branch did protect the rights of the bungalow owners, it failed to preserve the integrity of the neighborhood. It is hard to imagine a future other than the gradual replacement of the few remaining modest bungalows in Long Branch and elsewhere along the Shore through attrition, particularly when owners of modest means must face increased hazard insurance.

GENTRIFICATION. Unlike redevelopment, which removes older structures to make way for new ones, gentrification preserves and renovates the built environment but typically replaces low-income (and often minority) residents with

3.3. Beachfront North house for sale, Long Branch, winter 2014. Photo by author.

more affluent (and often white) ones. Gentrifiers are typically drawn from urban professional classes, who are often young and childless, with relatively high disposable incomes, allowing them to support trendy businesses like bars, art studios, and boutiques. Gentrification is often hotly contested, but it is hard for city governments to discourage changes that increase property and sales taxes and bring in affluent, childless residents who generally make few demands on the social service infrastructure that local government provides. Consequently, and following the advice of urban planner Richard Florida,[48] cities across North America have sought to attract the young, affluent "creative class" with restaurants, bars, and pedestrian-friendly consumer zones. To identify these economically desirable residents, Florida defines the creative class using a variety of measures, including concentrations of high-tech professionals, artists/musicians, and gays and lesbians. Low-income residents can rarely resist the onslaught, politically organizing for a time but eventually ceding physical space and then entire neighborhoods to newcomers with greater economic resources and political clout.

In her study of the gentrification of New York neighborhoods, sociologist Sharon Zukin identified a pattern that applies to gentrifying areas along the Jersey Shore.[49] First, there must be an attractive but rundown housing stock available at relatively low prices, where the first wave of gentrifiers—often artists priced out of more central metropolitan neighborhoods and looking for cheap studio space—can gain a foothold. Pioneers develop and support a number of important social, economic, and cultural nodes in the gentrifying neighborhood, such as bars, studios, and galleries, but these generally coexist with existing neighborhood businesses. As amenities catering to the pioneers proliferate, they draw in a second wave of gentrifiers, who are more risk-averse but come with greater economic resources. Zukin observes that in a quest for novel and authentic experiences, this second wave of gentrifiers is particularly attracted to neighborhoods with distinctive cultural and historic identities, although new residents may be oblivious to how their presence undermines and displaces the very people who forged those distinctive place identities. Businesses that spring up to cater to the second wave of gentrifiers replace the long-standing businesses that served the original population. Finally, gentrifying neighborhoods become popular among even more mainstream and risk-averse residents and visitors (what Zukin calls "bourgeois bohemians"), displacing not only the low-income original residents but often the cultural pioneers who began the gentrification process.

The most evident locus of gentrification along the Jersey Shore in the last two decades includes the neighborhoods surrounding Wesley Lake. On the south shore of Wesley Lake is Ocean Grove, part of Neptune Township, historically known as a Methodist retreat and for its Victorian architecture. On the north shore is the historic commercial district centered on Cookman Avenue in Asbury Park. The two neighborhoods are joined by a stretch of boardwalk along the beach near the decaying oceanfront Asbury Park Casino. By the time of the Asbury Park riots in July 1970, both neighborhoods were on a downward trend, with their turn-of-the-century facilities showing signs of age and neglect. The subsequent abandonment of the Cookman Avenue commercial district after the Asbury Park riot was not associated with a rapid decline in population as had occurred in places like Newark following the civil disturbances there in 1967, and Asbury Park maintained its population of around 17,000 inhabitants until the first decade of the twenty-first century.[50] However, this population grew progressively poorer and less white, with large numbers of blacks and Latinos.

More recently, gentrification has dramatically changed the demographic profile of the oceanfront areas in Asbury Park and Ocean Grove. In 2000, Asbury Park had a population of 16,930 individuals, of whom 62.1 percent were black/African American and 15.6 percent Latino. More than half of all households (53.1 percent) were "family households" defined by the U.S. Census as having dependent children, and only 19.5 percent of housing units were owner-occupied. Just ten years later, in 2010, population had dropped nearly 5 percent to 16,116, but whites had increased by 11.7 percent (from 24.8 percent in 2000 to 36.5 percent in 2010, and to 42.9 percent when only considering oceanfront census tracts), owner-occupied homes rose by 0.7 percent (from 19.5 percent to

3.4. Ocean Grove, Wesley Lake, and Cookman Avenue, Asbury Park, from above, winter 2014. Photo by author.

20.2 percent), and seasonal homes by 2.8 percent (from 0.6 percent to 3.4 percent). The city's black population declined in population over the same period, to 51.3 percent in 2010, with the oceanfront census tracts dropping to only 33.5 percent black. Likewise, family households declined to 47.2 percent in 2010, and only 38.7 percent in Asbury's two oceanfront census tracts.

In neighboring Ocean Grove, whites, homeowners, and seasonal homes also increased, even as blacks, families, and year-round residents declined. In 2000, Ocean Grove had 4,256 residents and was 93.1 percent white, with 33.7 percent families, and 19.1 percent of its housing stock seasonal. Ten years later, Ocean Grove's population had dropped by more than 20 percent to 3,342, and less than one-third of households (31.6 percent) were families. Ocean Grove got even whiter (to 98.6 percent white) and the number of seasonal homes grew to 29.5 percent of all housing units. As a point of comparison, Neptune Township as a whole (which contains Ocean Grove but also borders Asbury Park to the west) gained less than 1 percent population during this decade, with both white and black populations, percentage family households, and percentage owner-occupied units remaining more or less stable across the decade.

Asbury Park and Ocean Grove have experienced these demographic changes at the same time that they have largely preserved the historic nature of their homes and businesses. In 2010, only 5.8 percent of all housing units in Asbury Park had been built since 2000, and 37.6 percent of housing units were built before 1940. Preservation was much more notable in Ocean Grove, where a mere 1.2 percent of all housing had been built since 2000 and a full 65.9 percent had been built before 1940. The age of housing in both areas, when coupled with the demographic shift toward whites, homeowners, and second-home owners, and away from blacks and families, is a hallmark of gentrification. As a result, the values of homes have increased much faster during this decade in both locations than in other parts of the state. The median values of owner-occupied homes in the census tracts that include the Cookman Avenue section of Asbury Park and Ocean Grove have increased dramatically: in Asbury, the median value of owner-occupied units more than doubled from $127,100 in 2000 to $335,200 in 2010; in Ocean Grove, values rose from $149,200 to $390,000.

These gentrified communities are now marketing themselves as destinations attractive to members of Richard Florida's creative class. The Ocean Grove Chamber of Commerce bills itself as "A Timeless Treasure at the Jersey Shore"[51] and features recommendations by Fodor's travel guides and the Tripadvisor website in its "About Ocean Grove" section.[52] The website contains multiple photographs and references to its historic landmarks, societies, and national historic recognition. Asbury Park's 2013 City Guide refers to Asbury Park as a "small hip city and resort town" and the guide's very first page specifies that the downtown commercial area, while full of ethnically diverse eateries, features "retail shops and resources for home renovations."[53] In fact, the Asbury Park guide pretty much reads as if its authors had carefully studied Florida's works before deciding what to highlight.

Socially and economically, and consistent with Florida's use of same-sex households to measure the presence of the creative class, the most visible gentrifiers in Ocean Grove and along Cookman Avenue have been gay, lesbian, bisexual, queer, and transgendered (GLBQT) residents and business owners. A Williams Institute report on the 2010 census found that Ocean Grove had the highest concentration of same-sex households in New Jersey, followed by Lambertville and Asbury Park.[54] These concentrations place these Monmouth County Shore towns among those with the highest proportion of same-sex couples nationwide, with Ocean Grove ranking sixteenth and Asbury Park ranking nineteenth among all U.S. cities with fewer than 100,000 people, and behind only Provincetown, Massachusetts, and Rehoboth Beach, Delaware, among beach towns in the Northeast.[55] In both Asbury Park and Ocean Grove, concentrations of same-sex households (40.16 same-sex households per 1,000 households in Ocean Grove and 37.91 in Asbury Park) exceeded concentrations in all large and midsize U.S. cities, intensifying the social, cultural, and economic impact of the GLBQT population. Although the Human Rights Campaign gave both cities relatively low ratings on its "Municipal Equality Index" (Ocean Grove scored 77, while Asbury scored just 59 on a scale of 0–100),[56] the two municipalities are widely recognized as important centers for the GLBQT community in the Garden State and regionally, and GLBQT residents, business owners, and visitors compose an increasingly important segment of the region's place entrepreneurs.

Redevelopment of Resorts after Sandy

As noted earlier, New Jersey dodged an economic bullet when Sandy damaged the more affluent and resilient Monmouth and Ocean more heavily than Atlantic and Cape May Counties. Growth coalition interests from the public and private sectors quickly formed alliances to rebuild their common economic base and made resort areas a major focus of redevelopment efforts. Sometimes this common cause rebuilt fractured coalitions, such as in Point Pleasant Beach, where the local government accepted one million dollars from a bar owner to repair its boardwalk, even though the municipality had changed its laws to curb the rowdiness from that bar just the summer before.[57] Likewise, to receive federal grants for its repair, Ocean Grove's boardwalk had to be officially recognized as public property through an effort of state and local politicians and the Methodist Camp Association.[58] Just four years before, the association had lost a divisive battle against the state in the New Jersey Supreme Court about permitting same-sex civil union ceremonies in its Boardwalk Pavilion.[59] Asbury Park, whose boardwalk was damaged but not washed away in Sandy, was able to lure visitors to its gentrifying amenities—it was one of only a small number of towns that reported more beach badge sales in 2013 than in the year before.[60] Recovery aid was also a boon to Asbury Park, which secured $3.6 million dollars in federal grants to repair its boardwalk.[61]

In contrast, less affluent resort areas continue to suffer after Sandy. Seaside Heights had a devastating summer 2013 season; a *Star-Ledger* headline described the Seaside Heights boardwalk as "awfully empty," and the accompanying story quoted business owners and local officials indicating that business was off 20 to 40 percent from the year before.[62] The borough's 2013 budget included anticipated reductions in beach, parking, and municipal court revenues.[63] Court revenues alone were later reported to have declined to $230,000.[64] Residents spoke of a town dark and empty at night in the summer, particularly since many of the residential and rental properties in town and to the immediate north on the Barnegat Peninsula remained uninhabitable. The additional loss of boardwalk businesses following the September 2013 fire aggravated the damage. One resident asked on behalf of local business owners, "How many times can you recover?"

Second-Home Ownership as the Alternative to Resorts

While older resorts declined and were redeveloped or gentrified, or struggled, a growing number of people in the postwar period also used their middle-class affluence to buy their own personal piece of the Jersey Shore, and second-home ownership boomed (see map 2). Second homes allow people to access the recreational amenities of the beach and cooler climates in the summer, but they also serve as a symbol of class position. According to the 2000 Housing Census, there were 109,075 seasonally vacant housing units in Atlantic, Cape May, Monmouth, and Ocean Counties (see table 3.6). The highest concentrations of seasonally vacant homes stretch along barrier islands between the historic resorts of Atlantic City and Cape May, but seasonal home ownership is also prevalent on Long Beach Island (Long Beach, Surf City, and Beach Haven) and the Barnegat Peninsula (Toms River, Stafford, Brick, and Berkeley, all of which have substantial inland territories, so proportionately fewer seasonal homes than in municipalities entirely located on the barrier islands). Because they are concentrated on the barrier islands, generally distant from employment centers in New York, North Jersey, and Philadelphia, ownership of shore homes denotes an abundance of leisure time.

Second-home ownership may be viewed by some as an example of conspicuous consumption and gratuitous waste—owning a home that one does not even live in—even while providing opportunities to consume and spend leisure time. At the same time, the types of homes that were historically built on the

TABLE 3.6.

Jersey Shore Municipalities with the Largest Number of Seasonally Vacant Homes, in Descending Order

Municipality	*County*	*Number of seasonally vacant units*	*Total number of units*	*Percentage of total units seasonally vacant*
Ocean City	Atlantic	11,440	20,298	56
Long Beach	Ocean	6,132	9,023	68
Sea Isle City	Atlantic	4,864	6,622	73
North Wildwood	Atlantic	4,558	7,411	62
Toms River	Ocean	4,351	41,116	11
Lower Township	Cape May	4,115	13,924	30
Avalon	Cape May	3,697	5,281	70
Wildwood	Cape May	3,302	6,488	51
Brigantine	Atlantic	3,134	9,304	34
Wildwood Crest	Cape May	2,760	4,862	57
Stafford	Ocean	2,591	11,522	22
Margate	Atlantic	2,553	7,006	36
Stone Harbor	Atlantic	2,549	3,428	74
Brick	Ocean	2,137	32,689	7
Cape May City	Cape May	2,089	4,064	51
Atlantic City	Atlantic	1,945	20,219	10
Ventnor	Atlantic	1,870	8,009	23
Surf City	Ocean	1,649	2,621	63
Beach Haven	Ocean	1,542	2,555	60
Berkeley	Ocean	1,511	22,288	7

Source: U.S. Census Bureau, Census 2000.

Jersey Shore were small and modest, often clustered in fairly dense communities of similar small, modest second homes. Especially outside of traditionally elite locations like Mantoloking and Spring Lake, second homes on the Jersey Shore are likely the products of earned income and not hereditary wealth. Moreover, ownership of a Shore home often means considerable extra work and forgoing of other recreational activities or consumer goods. Nonetheless, Shore homes

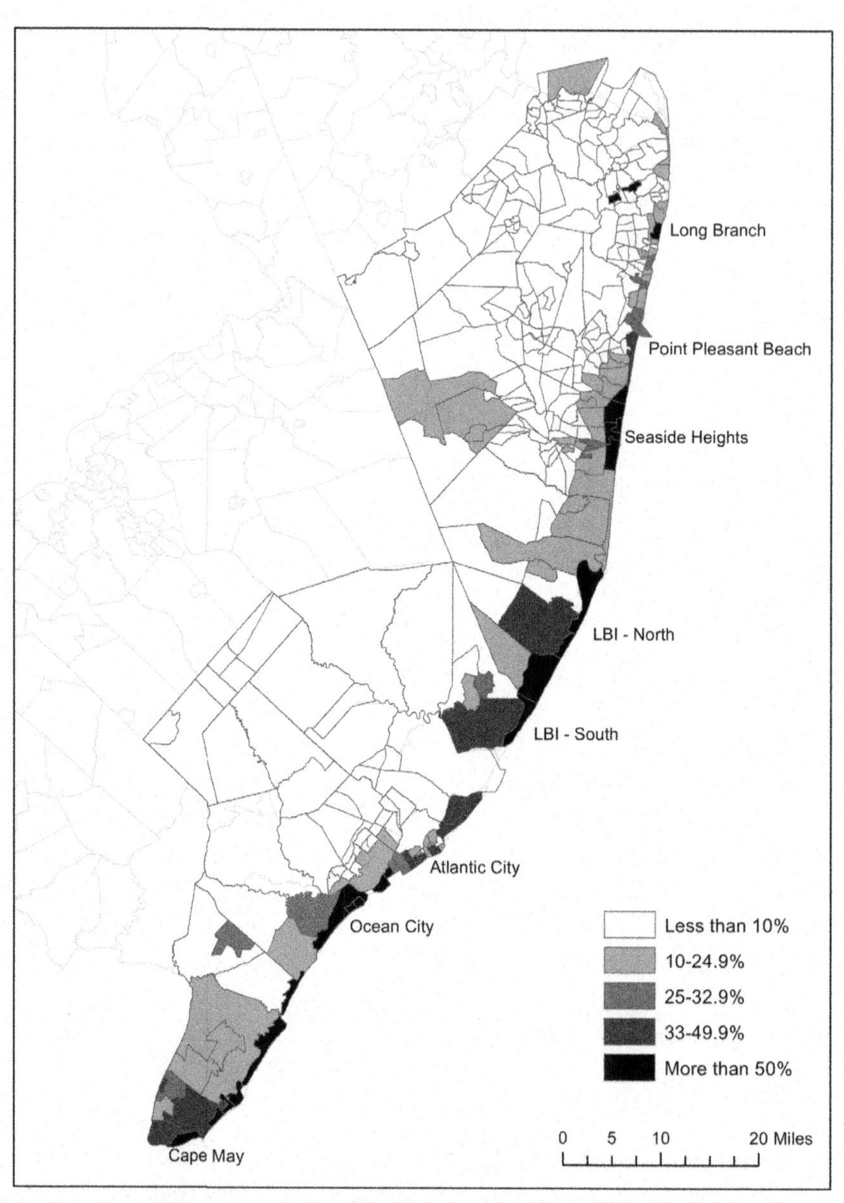

MAP 2. Seasonally vacant properties as percentage of all housing units. Map by author, based on GIS shapefiles from New Jersey Department of Environmental Protection, NJ-GeoWeb.

were and are viewed as desirable by wide segments of the working and middle classes in New Jersey and neighboring states, especially post–World War II boom years. Although segments of the working classes were excluded from the possibility of seasonal homeownership (most notably racial minorities, and especially African Americans), the Shore remained accessible for a variety of income groups.

This remains more or less true today; while there are certainly economically elite owners, second-home ownership on the Jersey shore does not necessarily confer the elite status of having, for example, a second home in the Caribbean or a pied-à-terre in Manhattan or Paris. Regionally, having a home on the Jersey Shore may confer less prestige than a house in the Hamptons, or even on the Outer Banks of North Carolina. A second home on the Jersey Shore is frequently seen as evidence of the hard-working but humble origins of today's upper and middle professional classes, rather than evidence of hereditary wealth. Shore homes are frequently passed across generations in families with origins in the white ethnic working classes of the early to middle twentieth century. As such, Shore homes are critical symbols of intergenerational mobility.

The destruction of Shore homes by Sandy threatens to erase that history, particularly if the cost of reconstruction and flood insurance goes beyond the affordability of families who still rely on income rather than wealth. The prospect of the Jersey Shore becoming less affordable for middle-income groups had started before Sandy struck. The National Flood Insurance Program had offered

3.5. Family reunion at a Long Beach Island shore house, Beach Haven, summer 2012. Photo by author.

subsidized flood insurance on roughly 12,300 secondary homes in New Jersey.[65] In July 2012, just three months before Sandy, it was amended by the Biggert-Waters Act to raise the cost of flood insurance premiums on second homes in order to cover debts incurred by the program related to Hurricane Katrina and other costly storms. The Biggert-Waters Act mandated that insurance premiums would increase up to 25 percent per year until the true cost of flood insurance was borne by the homeowner, although insurance costs could be reduced through elevating properties above base-flood elevations designated by Federal Emergency Management Agency (FEMA) maps.[66]

The post-Sandy release of new FEMA advisory maps of the Jersey Shore on December 15, 2012, led the *Star-Ledger* to predict in its headline: "Jersey Shore Revolution Begins."[67] Advisory maps placed large swaths of oceanfront, bay, and river communities in the highest risk "V (velocity) Zone," which would translate into expensive insurance premiums, elevation costs, or both. In Monmouth and Ocean Counties, more than 43,000 acres of property were added to the previous V Zone, which indicates a greater than 1 percent annual risk for wave action and flooding.[68] On June 17, 2013, the maps were replaced by more detailed maps that removed from the V Zone about half of that land, amounting to 2,698 acres in Monmouth County and 20,808 acres in Ocean County.[69] However, much of the land removed from the V Zone was redesignated as A Zone, which is still considered at "high risk" for flooding (1 percent annual chance). Both A and V Zones require elevation of structures above the FEMA-designated base flood elevation to receive subsidized insurance, with higher elevations translating directly into lower flood insurance rates. The cost of elevation is hardly incidental, however; one contractor estimated it to be between $45,000 and $50,000 for an average home on Long Beach Island.[70] With costs like these, it is reasonable to expect that second-home ownership will become increasingly expensive on the Jersey Shore.

This chapter has examined the development of resorts and second-home communities on the Jersey Shore as an outcome of processes consistent with Molotch's theory of the growth machine. Place entrepreneurs converted elite resorts to destinations to cater to the rising incomes of the region's middle classes. In the prewar period, economic development emphasized dense resort communities like Atlantic City and Asbury Park. In the postwar period, relatively modest second homes proliferated in waterfront communities even as seaside resorts fell into decline. Regional economic growth in the 1990s has contributed to the redevelopment of certain parts of the Shore, threatening to replace both the poor in older resort areas and modest homeowners in second-home communities through rising property values, gentrification, and eminent domain. Damage from Sandy will likely intensify the progressive unaffordability of Shore communities, both through direct displacement and as flood insurance and mitigation costs make second homes less accessible to people with moderate incomes.

While considering the effects of Sandy on resorts and second-home communities, it is also important to remember that many thousands of people live at the Shore all year. The next chapter addresses year-round populations in New Jersey's four coastal counties. Year-round residents of the Jersey Shore view their home region differently than do tourists, second-home owners, and those who depend economically on the tourism industry. Year-round residents, or Shore locals, are whiter, less educated, and less affluent that the state's residents on average. Chapter 4 turns from examining how developers have reshaped the Shore to how class and racial dynamics have created a region of environmentally privileged and disadvantaged communities.

4

The Suburban Shore

Although the Jersey Shore is known for its tourism, most of the region's population lives there year-round. More than half the homes in barrier island municipalities from Long Beach Island (LBI) south are seasonally vacant, although Atlantic City is a notable exception to this pattern. With a few exceptions, in Monmouth County and in Ocean County north of LBI, the vast majority of Shore residents are year-round residents. Although they live in an amenity-rich environment, year-round residents are rarely as affluent as seasonal second-home owners, and they are more likely to work outside the tourism sector. For example, in Atlantic and Cape May Counties, roughly half of the workforce is dependent on tourism, but a relatively small minority of Monmouth or Ocean County residents works in this sector (6.7 and 10.7 percent, respectively). These patterns are even stronger moving away from waterfront areas and may be strongest in the inland communities that are linked economically and culturally to the Jersey Shore. In truth, Monmouth and Ocean Counties are better typified as suburban hinterlands to the New York metropolitan region than as tourism-dependent economies. The Shore also attracts relatively large numbers of retirees, who populate large age-restricted residential complexes in places like Lakewood, Toms River, and Berkeley Township, or permanently occupy formerly seasonal homes in their retirement.

This chapter examines the year-round population of the Jersey Shore, people whose lives, not just vacations, were directly disrupted by Sandy. To do so, this chapter must also examine class dynamics, with attention to how class intersects with race, ethnicity, and environmental quality. Through the middle of the twentieth century, most of the Jersey Shore was rural and remote. Monmouth County had been a colonial destination for farmers, and later industrial development in the Raritan Valley and in places like Red Bank translated into larger populations in this county. In Ocean, Atlantic, and Cape May Counties, populations remained sparse until the latter decades of the twentieth century; even today, many inland areas remain more forest than subdivision (see map 3). However, suburban developments since 1980, mainly designed for middle-class and predominantly white residents, have dramatically changed the landscape of the Shore and put hundreds of thousands more people at risk in major coastal storms like Sandy. Not surprisingly, these predominantly white, middle-class residents have responded to Sandy in ways that are predictable given their social position. This chapter examines the explosive suburban growth associated with these two northern-most coastal counties as a window to examining how suburbanization on the Shore is a mechanism for reproducing spatial exclusion by race and class, and how this segregation may create or sustain environmental privilege in the Sandy recovery period.

A Demographic Profile of the Shore in 2012

In spite of the relatively high value of waterfront property, the Jersey Shore region is actually less affluent than other parts of the Garden State. The 2012 U.S. Census' American Community Survey (ACS) estimated the median household income in New Jersey to be $71,637 and the mean household income to be $96,602, which indicates a distinct skew toward more affluent households. Among Shore counties, only Monmouth County has means and medians that exceed the state averages (median of $84,748 and mean of $113,330). Ocean County's median household income is only 85 percent of the statewide average at $61,038, Cape May County's is only 79 percent at $56,370, and Atlantic County's median income is only about three-quarters of the state average at $56,370. At the same time, the Shore region has a smaller percentage (15.7 percent) of households earning less than $25,000 a year than the 17 percent of households statewide, even accounting for the fact that the Shore contains notably more households receiving Social Security income (34.8 percent compared to 28.3 percent statewide). In other words, while the Jersey Shore is not as affluent as the state as a whole, it also is not as poor. Instead, the Shore is a region where middle-income households predominate.

Income patterns are linked to differences in education levels and occupational sectors; here again, the American Community Survey finds notable differences between the Shore and other parts of the state. Monmouth County, which has the largest, most affluent, most diverse, and most unequal population of the four coastal counties, is the only one where the proportion of residents aged twenty-five and over are more likely than New Jersey residents overall to have earned at least a four-year college degree (40.3 percent in Monmouth, compared to 35.4 percent statewide). Considerably smaller proportions of Cape

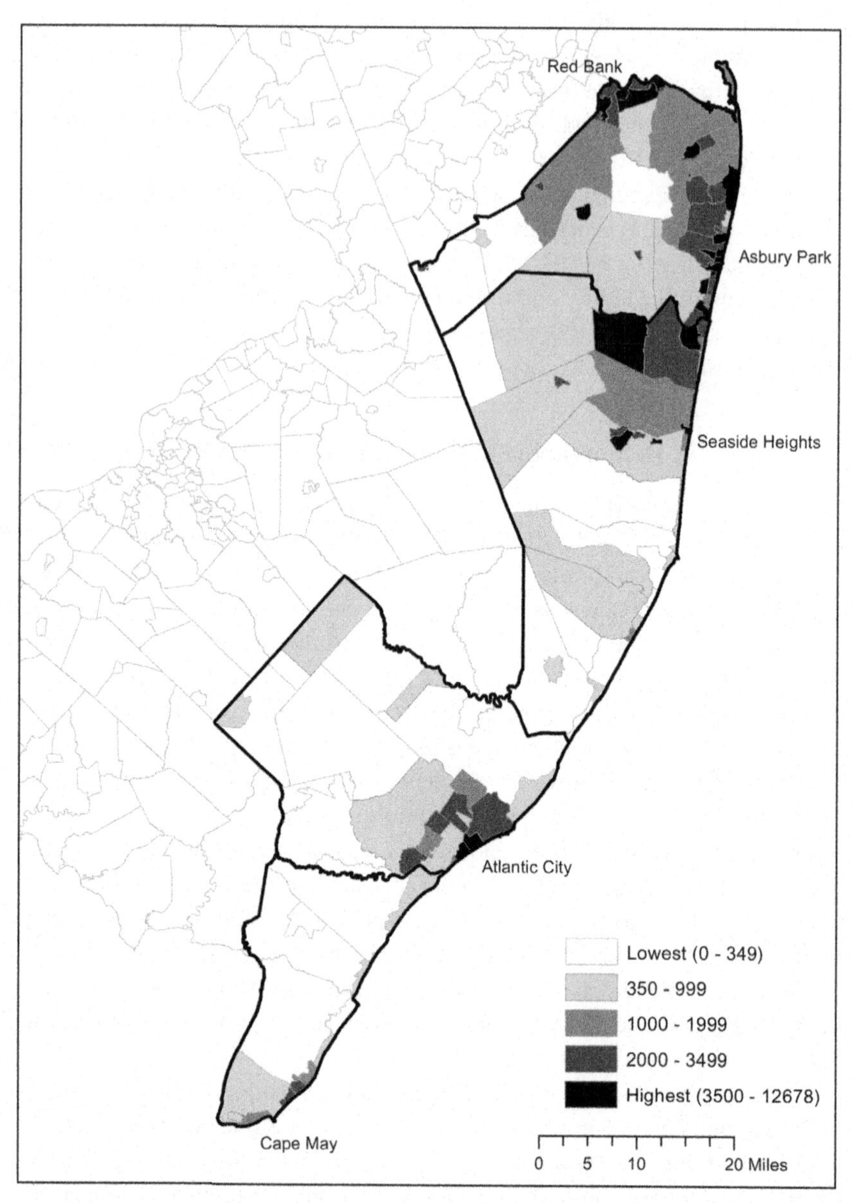

MAP 3. Population density of municipalities in Shore counties (people per square mile). Map by author, based on GIS shapefiles from New Jersey Department of Environmental Protection, "NJ-GeoWeb."

May (28.8 percent), Ocean (25.0 percent), and Atlantic (24.1 percent) County residents had earned college degrees. Also, only in Monmouth County were residents aged twenty to twenty-four more likely to be enrolled in college than statewide, with 46.9 percent in Monmouth compared to 45.4 percent in New Jersey. This is also in contrast to 41.5 percent in Ocean, 32.8 percent in Cape May, and only 16.8 percent in Atlantic County.[1] In terms of occupations, Monmouth County has slightly more civilian employed workers aged sixteen or over in the lucrative occupational sectors, such as information, FIRE (financial, insurance, and real estate), and professional or scientific management, with 26.9 percent of workers in these jobs, compared to 24.4 percent statewide, and 18.6 percent in Ocean, 16.4 percent in Cape May, and only 14.0 percent in Atlantic County.

The Shore is also notably more white than the state as a whole (see map 4), and is also more native-born. The 2010 Census recorded that in New Jersey, 17.7 percent of residents were Latino, 14.8 percent were black or African American, and 9.1 percent were Asian or Pacific Islander. In addition, more than one in five residents of the Garden State (20.8 percent) was born in another country. In contrast, the populations of the four Shore counties include only 8.0 percent black/African American, 4.6 percent Asian or Pacific Islander, and 10.2 percent Latino. Only 11.3 percent were born abroad. Based on the ACS in 2012, the Census Bureau classifies more than 90 percent of Ocean and Cape May County residents as "One race, white alone" (91.5 percent and 90.8 percent, respectively); 83.5 percent of Monmouth County residents were also so classified, compared 69.6 percent statewide. Only Atlantic County, with its concentration of nonwhites and Latinos in the Atlantic City metropolitan area, demonstrated greater diversity, with 65.9 percent of its residents reported as "One race, white alone."

These census data suggest that residents of the Shore counties, especially outside the Atlantic City region, can be described as white middle- and working class. Monmouth County is an exception to this pattern, mainly as a result of its proximity to employment centers in North Jersey and direct ferry connections to Manhattan, which increases the number of commuters; in fact, Monmouth workers have the longest average commute time of the four counties. Countywide figures, however, mask important differences within the Shore region about the distribution of residents; census tracts provide a more nuanced description of the Shore's residents.

Not surprisingly, waterfront census tracts are whiter and more affluent than the Shore region as a whole, but the whites living in these tracts are not necessarily elite populations. In oceanfront census tracts, 85.7 percent of residents of the Shore are whites, and this percentage rises to 89.3 percent when oceanfront tracts in Atlantic City are not counted. This racial pattern persists when including all waterfront census tracts (ocean, river, and bayfront): 86.07 percent of waterfront residents are white, rising to 87.17 percent when Atlantic City is

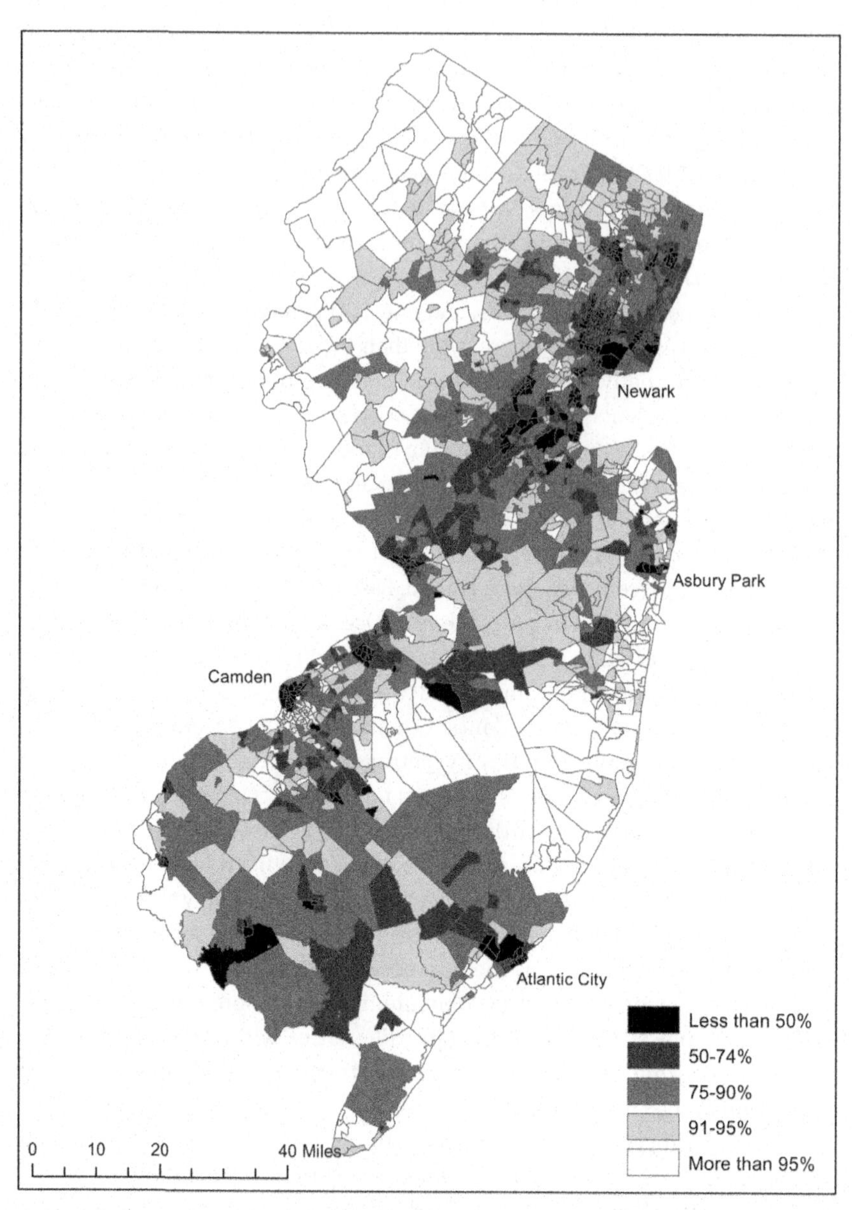

MAP 4. Whites as a percentage of total population by census tract in New Jersey. Map by author, based on GIS shapefiles from New Jersey Department of Environmental Protection, "NJ-GeoWeb."

excluded. White does not necessarily connote wealthy populations: the average median income of oceanfront tracts in 2012 was $61,699 ($65,078 excluding Atlantic City) and for waterfront tracts was $65,078. These are all *lower* than the state median ($71,637), which includes a much more racially and ethnically diverse population. The Raritan Bay shore census tracts in Monmouth County are examples of a white nonelite Shore region. Along the Bayshore waterfront tracts, 89.7 percent of residents are white, but the average median income is $63,154 and only 32.6 percent of residents over age twenty-five have earned a college degree. These figures drop to $56,139 and 23.6 percent when the gentrifying communities of Highlands and Atlantic Highlands are excluded.

This is not to suggest that race and class are not connected at the Shore. Communities with larger nonwhite populations are, on average, poorer than their white neighbors; the point is that most resident whites are not wealthy. Among all census tracts in the four Shore counties, the top twenty in terms of median household income average a median income of $139,998 per year and have residents who are 91.1 percent white and 87.3 percent college-educated. These tracts are spread throughout the four-county region, although concentrated in Monmouth County (see table 4.1). In contrast, the twenty census tracts with the lowest median income have an average median income of just $26,717 per year, and are composed of only 51.9 percent whites and only 18.9 percent college graduates (see table 4.2). Minority-majority census tracts in Asbury Park, Atlantic City, and Long Branch have some of the lowest average incomes on the Shore, but so do the mostly white retirement communities of Ocean County and the Orthodox Jewish residents of Lakewood.

There is little mystery about why more affluent communities are whiter, but the diverse composition of the middle and working classes of the Jersey Shore is less well understood. Although the Jersey Shore is hardly an elite region, socioeconomic class, race, and ethnicity clearly shape where people live, both in the state of New Jersey and within Shore counties. As residence has been conclusively linked to a variety of social outcomes, including early exposure to violence, educational attainment, and quality of life, how and why people arrange themselves residentially on the Jersey Shore merits further consideration. These class and race dynamics also provide insight into why the post-Sandy recovery efforts take the shape that they do.

The Benny Homeland

As much as the locals in Monmouth and Ocean Counties may hate to admit it, the region's local population is closely tied to patterns of population growth and redistribution in the New York City/North Jersey metropolitan region. Shore locals typically refer to summer visitors from this region disparagingly as

TABLE 4.1.

Highest-Income Census Tracts of the Jersey Shore

Tract number	*Municipality*	*County*	*Median income (dollars)*	*% white, non-Hispanic*
CT8038	Rumson	Monmouth	219,500	96
CT8033	Holmdel	Monmouth	180,234	80
CT8042	Little Silver	Monmouth	163,382	90
CT8032.02	Holmdel	Monmouth	161,417	84
CT8099.02	Colts Neck	Monmouth	159,635	94
CT8097.04	Marlboro	Monmouth	147,222	81
CT8012	Middletown	Monmouth	146,944	92
CT8099.01	Colts Neck	Monmouth	141,111	91
CT8046	Tinton Falls	Monmouth	130,865	84
CT8065.03	Ocean	Monmouth	130,743	87
CT8011	Middletown	Monmouth	130,156	91
CT8037	Fair Haven	Monmouth	126,929	93
CT0126.02	Linwood	Atlantic	124,375	89
CT8125.01	Millstone	Monmouth	123,008	85
CT7223	Toms River	Ocean	121,429	89
CT8015	Middletown	Monmouth	119,792	90
CT8101.01	Manalapan	Monmouth	119,420	92
CT8096	Marlboro	Monmouth	119,351	86
CT8119	Upper Freehold	Monmouth	117,257	89
CT8007.01	Middletown	Monmouth	117,188	94

Source: U.S. Census Bureau, Census 2000.

"Bennies," which appears to be derived from the New Jersey Shore train line's paper tickets, which included stations in Bayonne, Elizabeth, Newark, and New York.[2] These cities are undoubtedly part of the original urban core of New Jersey, including Hudson, Essex, and Union Counties. Along with southern Passaic and Bergen Counties, North Jersey was an important component of the industrial New York City region. At the turn of the twentieth century, the population in these five New Jersey counties represented more than one million people, with more than 78 percent of these in Hudson, Essex, and Union Counties. As a point of comparison, New York City in this same time period contained about

TABLE 4.2.

Lowest-Income Census Tracts of the Jersey Shore

Tract number	*Municipality*	*County*	*Median income (dollars)*	*% white, non-Hispanic*
CT0015	Atlantic City	Atlantic	14,246	15
CT0024	Atlantic City	Atlantic	18,358	29
CT0011	Atlantic City	Atlantic	20,129	6
CT8056	Long Branch	Monmouth	22,887	14
CT7153.02	Lakewood	Ocean	24,578	82
CT8073	Asbury Park	Monmouth	25,774	2
CT0014	Atlantic City	Atlantic	26,250	8
CT0019	Atlantic City	Atlantic	26,724	16
CT0025	Atlantic City	Atlantic	26,988	18
CT0005	Atlantic City	Atlantic	27,090	21
CT8076	Neptune	Monmouth	27,431	4
CT7154.02	Lakewood	Ocean	27,750	96
CT8072	Asbury Park	Monmouth	30,350	7
CT0023	Atlantic City	Atlantic	30,407	13
CT7159.022	Lakewood	Ocean	30,659	71
CT7312.02	Berkeley	Ocean	30,837	97
CT7201.02	Manchester	Ocean	30,865	96
CT0205	Woodbine	Cape May	30,927	43
CT0004	Atlantic City	Atlantic	30,938	35
CT7222	Toms River	Ocean	31,158	95

Source: U.S. Census Bureau, Census 2000.

3.5 million residents, including the recently consolidated outer boroughs of the Bronx, Brooklyn, Queens, and Staten Island.

North Jersey enjoyed the fruits of early industrialization in the New York metropolitan area, with the early development of factories, machines, and communication technologies. These benefited from infrastructural developments in ports, canals, and railroads, and the massive influx of white ethnic immigrants into the metropolitan region in the latter half of the nineteenth century, first from Ireland, and later from southern, central, and eastern Europe. When the National Origins Acts of 1921 and 1924 effectively shut off migration from southern, central, and eastern Europe, white ethnic immigrants were able to use their positions in the burgeoning industrial sectors to achieve upward class mobility. However, older industrial locations, especially those constrained by physical space, became obsolete as new assembly-line techniques lowered the cost of mass production but generally required larger single-story factories. North Jersey's older industrial facilities, typically housed in multifloor factories in a dense urban setting, were already showing decline on the eve of World War II, even though the five-county region doubled its population to more than two million inhabitants by this time.

The wartime growth of North Jersey industries, coupled with the shortage of immigrants as a result of both the war and xenophobic immigration restrictions, encouraged a Great Migration of African Americans from the Jim Crow South to urban centers in the North and West. In 1950, Newark had grown to its peak of 438,776 residents, of whom about 75,000 (about 17 percent) were African American. African Americans were often given the worst-paying and most dangerous jobs, but these were often better than opportunities that had been available in the South. Moreover, as upwardly mobile white ethnics relocated to the suburbs, subsidized by Veteran Housing Authority mortgages and the federal interstate highway system, they left behind relatively inexpensive housing in the urban cores. As a consequence, over the next two decades, more than 150,000 additional African Americans made Newark their home. This demographic shift led to tensions with the remaining white city population, eventually erupting into the 1967 Newark riots, which left more than one hundred dead and huge swaths of Newark's commercial districts in ashes.[3] The following year, the Federal Fair Housing Act removed racial restrictions from suburban housing, and the black middle class began to follow upwardly mobile whites into the suburbs. By 1970, Newark's population was 54.3 percent African American (table 4.3), but it had lost more than 55,000 residents. During the same period, Newark lost more than half of its jobs in manufacturing (see table 4.4), a decimation of its employment sector from which the city still has not recovered.

Although Newark is the starkest example, white flight occurred during the same time period in most of New Jersey's industrial cities, including Camden, Elizabeth, Jersey City, Trenton, and Paterson. In 2010, all had black populations higher—and in some cases, much higher—than the state's average of 14.8 percent (see table 4.5). White (and black middle-class) flight has reduced the size of urban populations from their peaks in the postwar period (see table 4.6). In some cities, vacant housing has been filled by immigrants, including large numbers of Puerto Ricans and Latin American immigrants. Although there is considerable variation, cities became more Hispanic and contained more foreign-born residents than the state as a whole (17.7 percent Hispanic and 20.6 percent foreign born). The people moving into cities were notably poorer than those who had left; table 4.5 demonstrates that by 2010, New Jersey's cities

TABLE 4.3.

Non-White/Anglo Population in Newark, 1940–2010

Census year	*Total population*	*Percentage black*	*Percentage Hispanic*
1940	429,760	10.8*	
1950	438,776	17.2*	
1960	405,220	34.4*	
1970	381,930	54.3	7.2†
1980	329,248	57.3	18.6
1990	275,221	56.3	25.1
2000	273,546	51.9	29.5
2010	277,140	53.9‡	33.8§

Sources: 1940–2000 Population Data: New Jersey State Data Center, "New Jersey Population Trends, 1790 to 2000" (Trenton, NJ: New Jersey State Data Center, 2001), http://lwd.dol.state.nj.us/labor/lpa/census/2kpub/njsdcp3.pdf; 1940–1970 Percent Black and Hispanic: U.S. Census Bureau, *City and County Data Books* (Washington, DC: U.S. Census Bureau, 1944–2000), http://www2.1ib.virginia.edu/ccdb/; 1980–2000 Percent Black and Hispanic: U.S. Department of Housing and Urban Development, "State of the Cities Data Retrieval System," http://socds.huduser.org/Census/Census_Home.html; all 2010: U.S. Census Bureau, Census 2010.

*Black and other nonwhite.

†"Spanish heritage."

‡Black alone or in combination with one or more other races.

§"Hispanic or Latino, of any race."

had higher level of unemployment, larger proportions of families receiving food stamps (SNAP benefits), and much higher proportions of children living in poverty than state averages, which were 8.7 percent unemployment, 5.8 percent receiving SNAP benefits, 13.1 percent of children under eighteen living in poverty.

The combination of a demographic shift in New Jersey's cities from predominantly white and middle class to predominantly nonwhite and relatively poor and the loss of manufacturing jobs in the cities led to a condition that sociologist William Julius Wilson has called a concentration of poverty.[4] The prospects for these cities are bleak. They face declining tax bases just as their populations become needier; schools in particular face immediate and costly needs with higher proportions of free- and reduced-price-lunch students and students in need of special services, such as English as a Second Language (ESL)

TABLE 4.4.

Newark's Manufacturing Workforce

Year	*Number of workers*	*Percentage change*	*Cumulative percentage loss*
1939	56,382	—	
1947	73,605	+30.5	—
1954	66,653	-9.4	-9.4
1958	57,270	-14.1	-22.2
1963	50,038	-12.6	-32.0
1967	47,500	-5.1	-35.5
1972	33,800	-28.8	-54.1

Source: U.S. Census Bureau, City and County Data Books.

TABLE 4.5.

Demographic and Economic Characteristics of New Jersey Cities, 2010 (percentages by category)

Category	*Newark*	*Camden*	*Elizabeth*	*Jersey City*	*Paterson*	*Trenton*
Black	53.9	50.4	22.5	27.7	33.7	54.0
Hispanic	33.8	47.0	59.5	27.6	57.6	33.7
Foreign born	26.6	13.2	46.5	38.2	29.3	23.1
Unemployed	15.7	22.4	10.9	10.3	10.7	15.3
SNAP recipients	20.6	30.3	10.2	10.5	19.8	18.9
Children in poverty	37.0	52.7	26.3	25.3	38.2	37.6
Vacant	16.2	16.1	12.0	13.0	12.1	19.6
Renters	75.1	60.0	72.7	68.2	70.2	57.9

Source: U.S. Census Bureau, Census 2010.

instruction. Relatively high levels of rental properties provide little neighborhood stability, which, coupled with relatively high rates of vacant properties and overall fewer resources to apply toward maintenance and upkeep of housing leads to a decline in the appearance of neighborhoods. Vacancy rates in city housing range from six to nearly ten times higher than the state average (1.9 percent), while the percentage of housing occupied by renters is nearly

TABLE 4.6.

Population Change in New Jersey's Historic Cities, 1940–2010

	Newark	*Camden*	*Elizabeth*	*Jersey City*	*Paterson*	*Trenton*
1940	429,760	117,536	10,9912	301,173	139,656	124,697
1950	438,776	124,555	112,817	299,017	139,336	128,009
1960	405,220	117,159	107,698	276,101	143,663	114,167
1970	381,930	102,551	112,654	260,350	144,824	104,786
1980	329,248	84,910	106,201	223,532	137,970	92,124
1990	275,221	87,492	110,002	228,537	140,891	88,675
2000	273,546	79,905	120,568	240,055	149,222	85,403
2010	277,140	77,344	124,969	247,597	146,199	84,913
Percentage change, 1940–2010	-35.5	-35.1	+13.7	-17.8	+4.7	-31.9

Source: New Jersey State Data Center and U.S. Census Bureau, Census 2000 and 2010 Census.

double the state average (33.4 percent) in almost all cities (see table 4.5). The poor families that Elijah Anderson called "decent" increasingly retreat into their own homes or move away, ceding public spaces to people with more nefarious "street" orientations.[5] Businesses decline, or are replaced by liquor stores, check-cashing storefronts, and other businesses that cater to the needs of low-income communities but often attract unwanted elements.[6] The end result in New Jersey is that many of the state's historic urban cores are now socially and economically isolated from the suburban communities that surround them.

Even while New Jersey's cities have declined in population and wealth, its suburbs have grown and prospered. Counties in the state's urban core grew by nearly 30 percent between 1940 and 2010 (table 4.7). Growth was facilitated by the interstate highway system and other infrastructural investments, which reduced travel times for commuters working in central cities in New Jersey, New York City, and Philadelphia. Suburban and northern exurban counties grew fastest, more than tripling their populations during the same time period. Even rural southern New Jersey nearly doubled its population in these seventy years. These years mark an important shift in the way residents of the state have spatially organized themselves, with whiter and more affluent residents least likely to live in urban cores and most likely to live in

TABLE 4.7.

Population Growth in New Jersey Counties, 1940–2010

	1940	*1980*	*2010*	*% Growth*	*% White*	*% Children in poverty*
Urban Core	*2,580,122*	*3,106,832*	*3,336,130*	*29.3*	*57.9*	*18.8*
Camden	255,727	502,824	513,657		65.3	17.6
Essex	837,340	778,206	783,969		42.6	20.6
Hudson	652,040	553,099	634,266		54.0	22.2
Mercer	197,318	325,824	366,513		61.4	15.2
Passaic	309,353	453,060	501,226		62.6	23.2
Union	328,344	493,819	536,499		61.3	14.0
Suburbia	*996,077*	*2,784,254*	*3,267,716*	*228.1*	*73.4*	*6.8*
Bergen	409,646	825,380	905,116		71.9	6.5
Burlington	97,013	395,380	448,734		73.8	6.7
Gloucester	72,219	230,082	288,288		83.6	9.9
Middlesex	217,077	671,780	809,858		58.6	9.3
Morris	125,732	421,353	492,276		82.6	4.4
Somerset	74,390	240,279	323,444		70.1	4.0
Exurbia	*116,579*	*330,326*	*386,306*	*231.4*	*91.7*	*7.1*
Hunterdon	36,766	107,776	128,349		91.4	4.1
Sussex	29,632	130,943	149,265		93.5	6.3
Warren	50,181	91,607	108,692		90.3	11.0
South Jersey	*115,458*	*197,542*	*222,981*	*93.1*	*71.3*	*20.4*
Cumberland	73,184	132,866	156,898		62.7	24.2
Salem	42,274	64,676	66,083		79.8	16.6
Shore	*351,929*	*112,5596*	*157,8758*	*348.6*	*82.2*	*14.1*
Atlantic	124,066	194,119	274,546	121.3	65.4	17.9
Cape May	28,919	82,266	97,265	236.3	89.8	13.6
Monmouth	161,238	503,173	630,380	291.0	82.6	8.3
Ocean	37,706	346,038	576,567	1429.1	91.0	16.4

Source: New Jersey State Data Center and U.S. Census Bureau, 2010 Census.

the exurban periphery, with fairly predictable gradations of both poverty and racial/ethnic diversity between them.

The decline of New Jersey's cities was accompanied by a rise in work in the service industries. Office parks and retail complexes began to spring up at the intersections of major highways, creating what urban planner Joel Garreau called "edge cities," distinguished by regional shopping centers, clusters of office parks, oceans of parking lots, and a commuter-shed that crept outwards in all directions.[7] A decade later sociologist Robert E. Lang described their evolution into "edgeless cities" of linear office and retail space that sprawled out in all directions, without even needing to cluster.[8] As work geographically shifted from urban cores to edge cities to edgeless cities, commuting became easier from farther away from urban cores, suburban development mushroomed, and exurban areas transformed from rural hinterlands to a continuous, albeit leafy, residential zone for the most affluent suburbanites. Given this centripetal movement, even if there were no natural amenities to lure affluent whites away from the urban core, over this time period counties on the Shore would have grown rapidly and would have become more affluent and white.

The Shore did offer more than just suburbia and a refuge from declining urban areas. Its natural environment included both scenic landscapes and recreational opportunities along its oceans, bays, and rivers. Although some of the waterfront communities were reserved for economic elites, many others were not. For those who could not afford waterfront properties, inland residences offered convenient access to these amenities. The cultural and sentimental ties of upwardly mobile, white ethnic populations to the Shore contributed to the sense that the Shore was a good place for nonelite whites who were no longer interested in living in New Jersey's cities.

Middle-Class Consumer Culture and the Shore

Since sociology's inception as a discipline, sociologists have pondered the implications of living in a society where we are linked to others through dissimilarity rather than sameness. Emile Durkheim, generally regarded as the father of the discipline,[9] described a division of labor in which the complex needs of modern society could only be met by promoting occupational specialization. Thus, a farmer specializes in food production, plumbers specialize in plumbing, teachers in teaching, and so on. One outcome of this specialization is that people in different occupations live very different lives and thus develop different subcultural patterns that threaten to cleave the sense of belonging to the same society. Modern societies thus live in a tension whereby people are highly differentiated economically and thus highly dependent on others (most of whom are strangers), with whom we have less and less in common in terms of our daily lived experiences, values, and norms. The trick to a modern society, thus, is to identify and promote social values that resonate with very diverse social, economic, and cultural groups.

In the United States, these social values echo libertarian ideals (liberty, equality) but have been heavily influenced by what sociologist Max Weber called "the Protestant work ethic," which emphasizes the importance of hard work and demonstrates a prohibition against (or at least ambivalence toward) ostentatious displays of wealth. These values are often consolidated under the idea that the United States is fundamentally a meritocratic land of opportunity where anyone who works hard deserves a chance to succeed. People who do succeed are thus assumed to have achieved their position by their own hard work and are therefore legitimately entitled to enjoy the benefits associated with that position. The meritocratic assumption explains how the United States can continue to tolerate persistent inequality despite its commitment to equality as a core value. Since people believe inequality results from differential effort on the part of individuals, it is a socially legitimate and, therefore, tolerable inequality. American sociologists Kingsley Davis and Wilbert E. Moore went so far as to assert that this inequality benefits modern society, in that it provides incentives for individuals to assume difficult and risky jobs.[10]

A contrasting view of inequality emerges from the Marxist tradition in sociology, which highlights the importance of whether one is an owner or a waged laborer in a capitalist system. In this view, owners and workers are not part of a single system that benefits everyone, but have diametrically opposed interests. Owners seek to make as much profit as they can in economic activities; one way this can be achieved is through paying workers as little as they can. This obviously runs against the interest of workers, who try to secure the highest wage possible. Because there were more workers than owners, Marx believed that an alternative to this system of exploitation could be established when instead of competing with one another for low-wage positions, workers joined together to overthrow the system. It is important to note, however, that Marx devised this theory in the context of the European nineteenth century, when there was a sufficient surplus population not only to fill the factories of European cities but also to emigrate by the millions to other regions of the world, including the United States. This large population created downward pressure on wages in Europe, and thus, a disadvantageous position for workers in the European capitalist system, even while generating unprecedented wealth for owners.

The story of labor in the United States in the nineteenth century was exactly the opposite of what Marx observed in Europe: while there was a pool of surplus labor in Europe, there were widespread labor shortages in the United States. Labor shortages allowed workers flexibility in choosing which jobs they would seek and keep, so employers had to attract them with higher wages. Labor scarcity was further heightened in the United States by xenophobic restrictions on immigration and ongoing racial and gender segregation in most occupations, including factory work. When Henry Ford introduced the assembly line for manufacturing automobiles in 1913, he found he had to pay workers an exorbitant five dollars a day to keep them from looking for less-repetitive work. Higher wages had the side benefit of creating a new market for Ford's products, ushering in an era in which relatively well paid workers began to drive the economy through their consumption patterns. As unions further empowered workers vis-à-vis employers, especially in manufacturing, the purchasing power of the working class increased, a situation that continued more or less into the 1970s. The postwar era solidified the economic and cultural centrality of the American middle-class consumer.

As the working classes morphed into a more affluent middle class, they became increasingly engaged in social activities that marked them as upwardly mobile. Departing from the Marxist tradition of locating class position in the productive process, in 1899, a sociologist from Minnesota defined classes on the basis of consumption.[11] Thorstein Veblen wrote that a "leisure class" is released from the burdens of economic activity by merit of its own accumulated surplus, and instead spends its efforts cultivating what French social theorist Pierre Bourdieu would later call "cultural capital."[12] In so doing, the leisure class distinguishes itself by demonstrating a refinement in manners and taste, in terms of both behavior and consumption. At the end of the nineteenth century and beginning of the twentieth century, growing relative wages allowed American workers to accumulate enough surplus to begin to emulate the leisure class as a means of displaying their own upward mobility. This process was enhanced by the fact that Americans were likely to live among people who were unaware of their modest class origins and could thus be persuaded by public displays of wealth.

Veblen identified three types of public displays of wealth that were valued as hallmarks of higher social status: conspicuous leisure, conspicuous consumption, and conspicuous waste. Through these, newfound members of the American middle-income working class demonstrate their newfound social status. All three types are characterized by public display and positive social evaluation, as it is only through others' acknowledgment that these activities become useful for indicating class position. Conspicuous leisure involves a public demonstration of not working or not having to work. Socially, leisure allows an individual or household to demonstrate enough pecuniary surplus that time can be devoted to nonwork activities. Veblen noted that leisure can be experienced vicariously, for example, by having other members of the household—such as women and children—refrain from waged work.

Conspicuous consumption involves the process of consuming for the sake of display, rather than for real needs or even creature comforts. Conspicuous consumption is most clearly seen in the acquisition of positional goods, which have value mainly because of their scarcity. An example of the conspicuous consumption of a positional good is the tradition of diamond engagement rings; these goods are acquired for no utilitarian purpose but serve to display wealth. Conspicuous waste refers to acquisition of things for the purpose of not using them. Suburban front yards are an excellent example of conspicuous waste: land is acquired and generally manicured and landscaped, but is not actually used for much other than appearances, as opposed to back yards, which are often outdoor living and recreational spaces.

The halcyon period of large Jersey Shore resorts directly supported the social mobility of the growing white ethnic middle classes of the postwar period who sought to accentuate their new position within the American class structure. Jersey Shore tourism blossomed only as the new middle classes found they had sufficient leisure time to engage in vacations and recreation. In fact, the rise of mass tourism is one of the main distinctions between older resort communities like Cape May, which catered to elites, and turn-of-the-century seaside resorts like Atlantic City. Bryant Simon's excellent study highlights how Atlantic City, although marketed as a resort for elites, mainly catered to upwardly mobile white ethnics.[13] According to Simon, Atlantic City owed its popularity at the turn of the century to its ability to provide public venues for conspicuously displaying leisure, consumption, and waste. Women and children were often sent to the resort for entire seasons of leisure, with working fathers joining them for weekends or even weeklong vacations from work. The boardwalk itself was a prime location for people to show off their newly earned affluence by wearing the latest fashions, consuming luxury foods and spirits in restaurants and nightclubs, purchasing functionally useless souvenirs, and either promenading or hiring an African American to push them along in a wicker chair. Simon highlights the importance of the racial dynamic in the conspicuous consumption of both entertainment and wicker chair rides, in that it allowed white ethnics, whose own white privilege was far from secure, to assert their ability to control the labor of others on the basis of race.

However, Atlantic City was a victim of its own success. As access to Atlantic City became broader, first across class and ethnic lines and later even in terms of race, the resort community became less viable as a means for denoting middle-, much less upper-, class status. This is part of the "resort cycle" detailed in the previous chapter, whereby in early stages of resorts, facilities are newer and access is limited to more elite tourists, who can afford the time and money required to reach more remote locations and have a cultural preference for newer, more exclusive, and more unusual locations. As resorts mature, facilities age and access becomes broader, making the resorts less appealing to elite tourists, who move on to other destinations. This has historically led business owners to struggle to make up for lost elites by making the resort accessible to even larger numbers of people, leading to both declining facilities and declining desirability, thus making it even harder to attract tourists.

Much like the aging cities of industrial North Jersey, Atlantic City and other seaside amusement–oriented resorts in New Jersey were hurt by a cultural shift in the means through which the middle class displayed status. As access to and use of these tourist facilities became more democratic, they ceased to function as exclusive spaces designed to distinguish one class of people from another: *anyone* could go to Atlantic City. Instead, the middle class began to display wealth through the acquisition of single-family suburban homes. Suburban homes offer ample opportunities to conspicuously not engage in waged labor, to consume socially desirable consumer goods (including vehicles), and to waste financial resources on nonfunctional items, such as a front lawn or garden. These cultural shifts were supported institutionally through the construction of federal and state highways after World War II, mortgage benefits for veterans (although only white veterans were fully able to take advantage of such benefits until 1968), and tax incentives that promote homeownership. When suburban families did consume tourism for social purposes, they increasingly made use of locations that were defined by continued economic and social exclusion, such as private theme parks and more distant and therefore more expensive destinations, including Caribbean resorts, Europe, and national parks in the U.S West. This is not to say that people stopped going to resorts along the Jersey Shore, but that their reasons for doing so no longer included the ability to demonstrate social status.

Whereas a generation before Shore resorts had served as ideal places to display conspicuous leisure, consumption, and waste for the upwardly mobile working class, by midcentury a suburban residence at the Jersey Shore indicated middle-class status. A second (vacation or retirement) home on the Jersey Shore was a marker of success, as was a waterfront property or residence in a traditionally elite town. These cultural forces, coupled with the spatial reorganization of New Jersey's urban northern core, help explain the explosive growth of the Shore since the end of World War II. As table 4.7 shows, populations on the Shore more than tripled between 1940 and 2010. By far, the greatest population growth was in Ocean County, which grew from a tiny population of just 37,706 in 1940 to 576,567 in 2010. Because there are few places to gentrify in Ocean County, the population that could not afford to squeeze into the county's dense waterfront areas has spread into the vast Pinelands in suburban subdivisions that reflect class preferences and racial dynamics.

4.1. Boundary of Island Beach State Park and Berkeley Township on the Barnegat Peninsula from above, fall 2013. Photo by author.

Environmental Injustice and Privilege

As with class and race relations, sociologists have long been interested in the relationship between who people are and where they live. Influenced by a combination of purchasing power, social policy, and consumer preferences, Americans increasingly separate (or, more accurately, segregate) themselves residentially by race, ethnicity, and class. This process has evolved over time to a point where older, previously industrial cities are characterized by urban cores containing socially and spatially isolated poor minorities (and sometimes affluent gentrifiers), but are surrounded by more modest suburban areas, which are, in turn, surrounded by more affluent exurban areas.

As Veblen and Bourdieu observed, social class can be conceived and measured not only through economics but also as a set of structural and cultural patterns that shape how and where people spend their time, whether at home, at work, or during recreational time. As New Jersey's white middle classes earned upward mobility, they fanned out into residential suburbs within commuting distance of original urban cores in North Jersey, Philadelphia, and New York City. Shifts in the racial, ethnic, and economic makeup of urban areas in the latter half of the twentieth century hastened the process of white flight, not only because some whites are racist but because urban lifestyle amenities could not complete with the single-family homes, green spaces, recreational opportunities, and schools available in the suburbs. This process was compounded by deindustrialization of New Jersey's economy and both the sectoral shift toward services and a spatial shift toward employment in the suburbs. As good jobs moved out of urban cores, transportation infrastructures grew and commuter-sheds expanded to include most of Monmouth County and northern Ocean County, which have taken on the characteristics of suburbs.

Where people choose to live is a combination of economic limitations and social and cultural preferences; most people will choose a home from among those they can afford in a location they select. Of course, this must also consider that historically, some communities could legally exclude those that they found undesirable. As a consequence, racial and ethnic minorities have been historically limited to certain geographic areas, often those with the worst housing stock and fewest environmental amenities. After the passage of the Fair Housing Act in 1968, some groups—especially African Americans—still face illegal housing discrimination. Princeton University sociologist Douglas Massey and his colleagues found that contemporary New Jersey is highly segregated by both class and race, with both elite groups and the urban poor minority populations living in more isolated locations than other groups.[14] This segregation is often reinforced by the political balkanization of New Jersey into 565 municipalities. Municipal codes can and have been used to restrict the type of housing that would be accessible to people of modest means.[15]

Apart from what residential land costs, there are important differences between its value when it is bought and sold (its "exchange value") and its value as a place to live (its "use value"). In *Urban Fortunes: The Political Economy of Place* (1987), sociologists John Logan and Harvey Molotch recognized that residential land represents a special type of commodity because of its unique combination of exchange and use values.[16] Middle-class homeowners, whose most important economic assets are usually the residences they occupy, undoubtedly have an economic interest in increasing the value of those residences and therefore they play an important role in growth coalitions. However, their economic interests are heavily influenced by and often subordinated to the "idiosyncratic locational benefits" of where they live, including the social and emotional ties that people have with specific places. People regularly choose occupations and individual jobs based on proximity to family, friends, and familiar landscapes, and they remain rooted to place in a manner that defies economic rationality. A 2012 report from the U.S. Census found that the proportion of Americans who have moved within the past five years has declined since 1970, and among those who have moved, more than half remain within the same county as their previous residence.[17] Moving was least common in the northeastern region of the United States, where more than 90 percent of individuals remained within the same county, despite significantly smaller geographic counties and the nearly built-out situation of the Boston–Washington corridor. For those who left their county of previous residence, the most common distance moved was fifty miles or less.[18] These figures underscore that the relationship people have with the places in which they live cannot be reduced to strictly economic terms.

The combination of affordability, social exclusion, and consumer preferences has created predictable and easily identifiable patterns in the environments in which people live. Poor, minority, and immigrant populations are more likely to be exposed to synthetic hazards, such as industrial pollutants and residential lead, as a result of their concentration in older, formerly industrial cities. Affluent residents enjoy safer and more aesthetically pleasing environments, although exurban regions may face heightened risk of natural hazards, such as storm surges along the Jersey Shore. These patterns of exposure to environmental hazards and amenities represent another major area of research, "environmental justice," which in sociology stems originally from seminal works by Robert Bullard and Paul Mohai.

While attending graduate school, urban sociologist Robert Bullard worked for an attorney who was hired by a group of upwardly mobile African American residents in suburban Houston, who discovered that the new facility being built in their community was to be a landfill. Bullard's subsequent research on where landfills and other toxic facilities were located in Houston uncovered an indisputable pattern of racial and ethnic bias.[19] Regardless of economic class, black and Latino neighborhoods shouldered a disproportionate burden of such facilities. Bullard then explored these patterns elsewhere in the southern states, culminating in his seminal study, *Dumping in Dixie: Race, Class, and Environmental Quality* (1990). In this book, Bullard uses "environmental racism" as a description of the differential exposure to polluted environments, rather than as an accusation of racist intent on the part of the people who decide where to site and operate toxic facilities. To Bullard, it does not matter if facilities' owners and operators are racist; what matters is that where these facilities are located translates into greater exposure to industrial contaminants for minority communities.

Just two years after the publication of *Dumping in Dixie*, University of Michigan sociologist Paul Mohai and his colleague in the School of Natural Resources and Environment, Bunyan Bryant, published an article titled "Environmental Injustice" and edited a book called *Race and the Incidence of Environmental Hazards*.[20] These publications identify both race and class as important in predicting exposure to pollution, with nonwhites and poor populations facing greater environmental risk. Mohai and Bryant demonstrate that while minority status or poverty alone predicts greater exposure to contaminated environments, poor minority communities are the most at risk. Mohai and Bryant joined Bullard and a burgeoning social movement in seeking "environmental justice" for these disadvantaged groups.

Since this time, environmental justice has become an important concern for both sociologists and policy makers. In 1994, President Bill Clinton signed Executive Order 12898 mandating that seventeen federal agencies track and affirmatively respond to how environmental risks are distributed across racial, ethnic, and economic groups.[21] A search for peer-reviewed articles on "environmental justice" in the Sociological Abstracts index of academic resources in 2014 produced 1,385 citations, indicating great academic interest on this subject since the identification of these patterns of exposure in the late 1980s.

More recently, University of Minnesota sociologists Lisa Sun-Hee Park and David Naguib Pellow have also explored the ability of affluent exurbanites to use their social position to secure even greater exclusive access to environmental amenities as an example of "environmental privilege." Park and Pellow investigated the elite resort community of Aspen, Colorado, where wealthy residents, including many who are vacation-home owners, have been able to dominate the local political structure. The elites point to their practices of green consumption (for example, of hybrid vehicles and organic foods), participation in elite seminars on environmental thought, and lengthy recreational times in the outdoors (skiing in the winter and hiking in the summer) to assert that they are best able to protect the stunning and unique environment of the Roaring Fork Valley. They also define the growing number of working-class Latinos in the area as a direct threat to the local environment. Park and Pellow acknowledge that the Latino workers of the region congregate in aesthetically unappealing trailer parks and other dilapidated residential areas, fail to spend a great deal of leisure time in the outdoors, and generally cannot afford to follow green consumption patterns. Park and Pellow also point out that by the nature of their humble patterns of consumption, the working class residents of the Roaring Fork Valley undoubtedly have less environmental impact than their mansion-dwelling, multiple-vehicle-owning, jet-setting neighbors, but they lack the political and cultural power to assert those claims in local politics. As a consequence, Latinos are progressively forced farther and farther down the Valley, often having to commute an hour in each direction to provide landscaping, housekeeping, and other services to the "green" elites of Aspen.

Environment privilege is the other side of the proverbial coin to environmental racism and classism, whereby undesirable and toxic facilities are differentially located in neighborhoods with poorer and predominantly black and Latino populations. Park and Pellow explain:

> Environmental privilege results from the exercise of economic, political, and cultural power that some groups enjoy, which enables them exclusive access to coveted environmental amenities such as forests, parks, mountains, rivers, coastal property, open lands, and elite neighborhoods. Environmental privilege is embodied in the fact that some groups

> can access spaces and resources, which are protected from the kinds of ecological harm that other groups are forced to contend with every day.[22]

Park and Pellow take great pains to point out that people who protect their environmental privilege are often well intentioned and see themselves as political progressives. They see themselves as worthy stewards of the natural environment, using environmental consciousness as a shield against accusations of elitism. In the same vein, people can justify their environmental privilege on the basis of their love for nature, as evidenced by their consumer and recreational patterns. Environmental privilege becomes regressive and racist, however, when these same people fail to reflect on their own contributions to environmental degradation and instead focus on less powerful groups as the source of environmental degradation. They also manipulate their greater access to political power to exclude nonwhites or less-affluent groups from having the same access to that environment. As Bullard pointed out, however, environmental injustice does not require the intent to discriminate, just a discriminatory outcome.

Some of New Jersey's waterfront communities are much like Aspen and other communities with environmental privilege, increasingly composed of elite residents and second-home owners. Some have been associated with wealth and affluence for much of their existence. In Monmouth County, Rumson, Sea Girt, and Spring Lake have long been retreats for economic elites, as have Ocean County communities like Mantoloking and Harvey Cedars. In general, these towns were settled early, have large lot sizes and homes, and relatively small populations. New Jersey law requires that even the most elite beach towns allow public access to the ocean, and the public is also allowed to use any part of the beach to the high-tide line. Beach towns may charge to use the beach facilities, with badges available usually for daily, weekly, or seasonal beach access. Elite beach towns follow the letter of the law but otherwise make it fairly difficult for nonresidents to use the beach through restricted parking and a lack of available public toilets and changing rooms. Mantoloking allows nonresident parking only on certain streets, and then only for two hours; they stress to nonresidents: "**If you plan to stay more than two hours** . . . there are no restrooms, places to eat, or shady areas, and on-street parking is limited" (bold in original).[23]

Mantoloking was one of the municipalities most affected by Sandy, but it is also one of the most affluent communities on the Shore. The 2010 Census recorded 535 housing units in Mantoloking Borough; of these 162 were occupied by year-round residents.[24] More than 80 percent of homes had values that exceeded one million dollars. When Sandy struck, 68 percent of the homes suffered major or severe damage, and every single residence had at least some damage caused by the storm.[25] The American Community Survey in 2013 reported only 468 housing units—reflecting the loss of the homes that disappeared in the storm and those that were immediately demolished. However, because nearly 70 percent of homes were already vacant for the season, the owners were safe in other locations. In other words, Mantoloking could wait. The small number of year-round residents dedicated their efforts to getting back into their own homes, securing the aid that they could, and planning an orderly recovery effort. The police organized themselves and the National Guard quickly to keep looters out of vacant, damaged homes. The borough was able to delay demolition until late spring 2013, when it successfully negotiated a single contract to remove the additional 50 homes that were damaged beyond repair. Two years after the storm, many lots in Mantoloking remain empty, and many other homes are under construction, but recovery is proceeding in an orderly fashion. Undoubtedly, Mantoloking will return to its elite Shore town status over time, with the same type of people and the same exclusive access.

In contrast, other communities that were devastated by Sandy have had tougher times. Some waterfront communities, particularly along the Raritan Bay shore and the Toms River beaches, are not particularly affluent. For example, the Bayshore's Union Beach in 2012 was a stable community of white working-class homeowners: 95.8 percent were white, and residents earned a median income of $66,419, which is slightly below the state median income. Most homes (84.8 percent) were occupied by a resident homeowner, with a modest median value of $259,300; New Jersey's median home value was $327,100. Only 18 of 2,269 housing units in Union Beach were seasonal homes, reflecting the fact that Union Beach is fundamentally a year-round community. Residents include a concentration of workers in skilled manual trades and a relatively large retired population (28.2 percent of households in 2012 received Social Security income). Similarly, Ortley Beach was made up of predominantly white (98.8 percent) homeowners (76.4 percent), although nearly 72 percent of the 2,658 housing units were seasonally vacant in 2010.[26] However, there were 498 owner-occupied homes in Ortley Beach. Year-round residents had a low median income in 2012 ($48,772), largely explained by the high proportion of retirees: a full 57.5 percent of households received Social Security. Although both communities were predominantly white, and neither is poor, they are not affluent in the same way that Mantoloking is.

The experience of less-affluent communities like Union Beach and Ortley Beach contrasts greatly with that of Mantoloking. Although relatively few residents rode out the storm in their homes and were spared that trauma, most also returned to the area as soon as possible, only to find that their homes, belongings, and neighborhoods were destroyed. Year-round residents who had access to their homes immediately after the storm began demolition and repairs at once, drawing on themselves, family, and an impressive army of volunteer laborers. Both residents and volunteers involved in this effort speak emotionally about the cleanup efforts, because people were making quick decisions to destroy and throw out furniture, appliances, personal items, carpets, and drywall in a state of shock and disbelief. Volunteers talk about piling personal belongings in veritable mountains along the roads, while homeowners sat and cried, unable to even participate in the process. They worked in cold and damp conditions, often without heat or electricity, being fed sandwiches by brigades of volunteers who drove around looking for people to feed. Louis Amaruso, the director of public works in Toms River, reported that in the thirty days after the storm, Toms River removed as much debris from the most affected areas—forty thousand tons—as the entire township had in the entire previous year.

Along the barrier islands in places like Ortley Beach, residents were unable to return to their homes for days and even weeks after the storm. They spent this time in hotels and shelters, or doubled up in the homes of friends or family, worried about their homes but unable to do anything about it. Toms River and other Shore towns organized buses to bring these residents to the damaged and often unsafe communities as soon as they were able, but residents were only allowed a suitcase or two to recover personal items. Some residents found their homes completely gone or utterly demolished when they arrived; others found disarray, mold, and choking smells. Mayor Thomas Kelaher of Toms River remembers, "I was there when the buses came back to see how people were doing. . . . People came back with empty

4.2. Ortley Beach from above, fall 2013. Photo by author.

suitcases because they couldn't find anything. One lady got off the bus clutching a ship's wheel with her kid's name on it, that's all she could find in what was left of her house."

Recovery was piecemeal—people did what they could with their homes with the resources they had available. Although volunteer brigades were eventually organized, most families initially relied entirely on themselves for demolition and cleanup, as access to the islands were restricted to homeowners to prevent looting. Residents recount the frustrating process of waiting to going through multiple armed National Guard checkpoints to reach their homes, then working only a couple of hours during the short winter days, because of both the lack of electricity and strictly enforced curfews. Others were overwhelmed by the extent of the damage and repairs. For example, a year after the storm, Mayor Kelaher reported that 956 demolition permits had been granted in Toms River, but only about half of the homes had actually been demolished. A surprising number of people in these communities decided not to rebuild. Many homes lacked sufficient money to replace what was lost, and others have decided that they would prefer to spend their time and money elsewhere. These homes remained vacant and unrepaired, becoming hazards, eyesores, and grim reminders for those who are trying to rebuild. Others have sold entirely, bringing new—and largely more affluent—homeowners into these same communities.

While the recovery experience for residents in places like Union Beach and Ortley Beach has been much more traumatic than in Mantoloking, middle-income residents have not been completely wiped out by the storm. Most of the volunteer and public sympathy and efforts in the recovery period were focused on their plight. People of modest means who had scrimped and saved to acquire their piece of the Jersey Shore were showcased and lionized; these were people who were deserving of public aid, charity, and goodwill. Even second-home owners earned sympathy, particularly those who had held properties for decades within families or had purchased Shore homes as part of their long-term retirement plans.

Residents of Union Beach, Ortley Beach, and other middle-class towns felt entitled to receive assistance from the government in this time of need and bristled when they were told that they were ineligible for many programs because they owned another property elsewhere or had incomes that were too high. They then used their resources and political power to effect changes in public policy. For example, the financial difficulties of Shore residents led to a reversal of the Biggert-Waters Act, which would have required homeowners in high flood-risk areas to pay insurance premiums that cover the actual cost of damage. The reversal of the act means that the cost of providing insurance in flood-prone areas will continue to be subsidized by American taxpayers far from the Jersey Shore. Few of these residents would see themselves as being privileged, particularly when they compare themselves to affluent second-home owners in places like Mantoloking. In fact, Point Pleasant Beach native Joanna Peluso found that middle-income homeowners used what she called an "appropriated environmental justice frame" to assert their claims to recovery resources; resident comments in public meetings demonstrated that residents believed that they were systematically deprived of their fair share of public funds.[27]

At the same time, the ability of middle-income homeowners to bend public policy to their interests is a form of environmental privilege. Dedicating resources to redeveloping communities devastated by the storm means that the federal and state governments cannot spend that money elsewhere on programs that might reach larger segments of the population. Billions in federal and state public funds have already been sunk into the reconstruction efforts in places that are destined to be damaged again in the future. Likewise, the focus on homeowners has marginalized the vulnerable renter population of the Shore. Renters were displaced along with homeowners, but rarely had access to insurance to recover losses, much less grants for rebuilding. In places like Seaside Heights, renters were permanently displaced without the ability to recover either belongings or the cash security deposits that might allow them to secure another apartment. The reduced number of available rental units in Monmouth and Ocean Counties and competition from displaced homeowners drove rental rates up in the storm's aftermath. As year-round renters are likely to become increasingly scarce in Shore towns, they also lose their ability to participate in the political process of recovery.

As the Sandy recovery experience demonstrates, the role of the government is central to understanding the relationship modern people have with their environments; consequently, the next chapter examines the role of government in more detail. That chapter considers the actions of government at many levels to determine how the state has or has not followed through on its duty to protect the Jersey Shore. The Federal Emergency Management Agency (FEMA) urges us to "Plan, Prepare, and Mitigate" for disasters, which is a nice slogan for a range of activities and responsibilities that span how individuals, organizations, and government at all levels should approach potential and existing troubles. Planning, preparing, and mitigating are also excellent ways to organize the ways in which public policy and agencies shaped the Jersey Shore both before and after Sandy.

5

Government, Bureaucracy, and Technical Fixes

Charles E. Fritz, one of the earliest sociologists who studied natural disasters, concluded in 1961 that natural disasters typically produce a strong sense of community in their immediate aftermath.[1] Fritz based his observation on a review of 161 disasters studied by social scientists, involving data collected from more than 20,000 individuals. He concluded that after the initial shock and concern about the welfare of loved ones passes, preexisting social categories, divisions, and allegiances are eroded by the focus on meeting the immediate needs of individuals who have survived the disaster. Fritz continued:

> People are able to perceive, with a clarity never before possible, a set of underlying basic values to which all people subscribe. They come to see that collective action is necessary for these values to be maintained. Individual and group goals and means become merged inextricably. . . . Thus, while the natural or human forces that created or precipitated the disaster appear hostile and punishing, the People who Survive become more friendly, sympathetic, and helpful than in normal times.[2]

Although a half-century old, Fritz's description generally fits the experiences of people who survived Sandy. Individuals were quick to help friends and neighbors. Volunteers from within Shore communities and from far, far away provided much-needed assistance in a variety of ways. For the most part, Shore residents described these experiences as strengthening their local communities and restoring their trust in humanity. In the immediate poststorm period, these warm feelings were extended to government agencies at all levels that rushed in to help. Over time, people began to express more frustration at government actors, particularly at the federal and state levels.

What is the role of the government in addressing storms like Sandy and their aftermath? The relationship between modern government and the everyday lives of people is poorly understood by most. After the massive bureaucratic expansion following World War II and into the 1970s, public agencies have increasingly faced a public hesitant to expand their activities and inclined to slash their budgets. Although some public officials, notably first responders like fire fighters, enjoy considerable goodwill among the public, regulatory agencies and elected officials have seen their public support wane. At the same time, when faced with disasters like Sandy, the public demands rapid response and attention, and is quick to accuse public actors of betrayal of public trust, often without considering whether actions are deliberately negligent and corrupt, or if public actors are simply limited by legal, organizational, and political constraints.

Home Rule and the Balkanization of the Shore

Perhaps the most important concept for understanding government at the local level in New Jersey is "home rule." Home rule refers to a set of relationships between local government (municipalities and counties) and state government in which local government has the legal authority to operate autonomously on most public matters. For example, home rule allows municipalities to elect their own form of local government, to create municipal offices and staff them as they see fit, to enter into reciprocal or fee-for-service contracts for a variety of public services, and to set local tax levels, including property and sales taxes. Real and perceived corruption in the state government, the lack of faith in "big government" in general, and a persistent belief in the superiority of participatory local government combine to contribute to the endurance of home rule in New Jersey. Home rule continues to hold significant political sway in a highly unequal and densely populated state, despite growing statewide regulation in the areas of environmental protection, affordable housing, and administrative/bureaucratic responsibility. In fact, proponents stemmed this erosion of home rule with an amendment to the New Jersey Constitution that requires all state mandates to local government to be funded by the state after 1996.[3] In other words, if the state wants to regulate something, it cannot burden local government with paying for compliance or enforcement.

As a result of the persistence of home rule, municipalities continue to have the legal authority to determine how land within the municipality is used, although they must make those decisions within regulatory frameworks set at the state and federal levels. For example, municipalities can decide which parts of their land should be dedicated for residential, commercial, or mixed uses. As long as the land is not protected by state or federal laws regarding environmental protection (such as the Endangered Species Act), as long as the proposed land use does not threaten to violate additional laws (such as the Clean Water Act), and as long as construction meets state codes (for example, uniform building codes), municipalities are largely free to make their own decisions about what gets built within their borders. In municipalities along the Shore, many municipalities must also work within regulations imposed by CAFRA (the Coastal Area Facilities Review Act) and the New Jersey Wetlands Protection Act, and may be subject to review by the Pinelands Commission. While this sounds like a lot of regulation, most land in municipalities remains unrestricted or faces limited regulation, not in the least part because much of the Jersey Shore was developed before these regulations were established, and virtually all have "grandfather" clauses that provide for leniency for properties that existed at the time these laws were enacted.

Home rule also allows for legal social sorting by residence, whereby wealthier and more politically powerful groups can pass local laws to limit interactions with less-powerful groups. As noted in the previous chapter, sociologist Douglas Massey and his coauthors have documented how New Jersey has become increasingly segregated by both class and race. Economic elites and the very poor are most likely to live isolated from other income groups, and both groups are becoming even more residentially segregated from the rest of society.[4] This is facilitated by home rule, which allows municipalities to enact exclusionary zoning laws that restrict the housing market to affluent buyers by requiring, for example, that all new properties have a minimum lot size or that prohibit multifamily units. The *Mount Laurel* decisions by the New Jersey Supreme Court do require that all municipalities provide for a "fair share" of regional affordable housing needs, but building affordable housing has been met with staunch opposition in many communities. People and municipalities that have resisted affordable housing quotas often invoke the sanctity of home rule. In an attempt to dismantle affordable housing obligations, Governor Chris Christie tried to disband the Council on Affordable Housing (COAH), the agency responsible for determining each municipality's fair share. When the New Jersey Supreme Court refused to let that happen, the administration produced new "fair share" quotas that allowed a sharp reduction in municipal obligations, but no explanation of how the new quotas had been determined.[5]

Home rule is particularly insidious in New Jersey because of three somewhat unusual features of the state's political organization. First, apart from federal properties (such as Gateway National Recreation Area and the Army–Air Force Joint Base McGuire-Dix-Lakehurst), all land in New Jersey is governed by a municipality; there are no unincorporated areas in the state. Second, there are a large number of municipalities in the state: 565, down from 566 when Princeton Borough and Princeton Township consolidated in 2013. Third, and not particularly surprising given the large number of municipalities, many of these municipalities are very small, both geographically and in terms of population. The four Shore counties contain a total of 125 municipalities, including a nest of micro-municipalities, with twenty-four occupying one mile square or less (see table 5.1 and map 5), an additional twenty-five occupying between one and two square miles, and another twenty-two occupying between three and five square miles. Only 40 of the 125 municipalities in the four coastal counties are larger than ten square miles, and most of these are well away from the water. Given the small geographical size of these municipalities, it is not surprising that small numbers of people live in them. For example, according to the 2010 census, Mantoloking in Ocean County had 296 residents, Cape May Point in Cape May County had 291 residents, and Loch Arbor in Monmouth County has just 194 residents.[6]

Small-population municipalities are, in theory, a good way to ensure participatory democracy. However, they create a variety of administrative difficulties, particularly when confronted with the specialized knowledge required to manage complex problems, including environmental and emergency management. In larger municipalities, a professional administrative staff manages most municipal functions and services. For example, the Shore's most populous waterfront municipality, Toms River, lists thirty-nine departments on its website, including police, fire, planning, emergency management, and health.[7] In contrast, most small municipalities in New Jersey have a commission-based administration or, more commonly, a small town council (with or without an elected mayor) that manages all functions of the municipality. Council members and commissioners, while probably accessible and familiar with their constituents, often do not work full time in their positions, nor do they have the specialized knowledge to plan for or mitigate disasters.

Only six of forty-five oceanfront communities are larger than ten square miles and only ten had populations greater than ten thousand people in 2010.[8] Thus, despite the complex interface of human habitation and coastal biogeography, Shore towns are frequently governed by part-time, nonspecialized municipal administrations. Public services are met through a variety of regional arrangements. This does not suggest in any way that these services are worse than if the municipality itself managed all these services, but it has created a convoluted set of responsibilities that make coordination complex. For example, while Loch Arbor oversees its own zoning and building codes, it contracts fire, first aid, and sanitation from Allenhurst, police from Deal, and health from a regional consortium located in Tinton Falls. Understandably, residents are not likely to be aware of which agencies (local or regional) are responsible for services until those services are needed, such as in the wake of a disaster like Sandy. Navigating these responsibilities while managing personal crises may

TABLE 5.1.

Smallest Municipalities in New Jersey Shore Counties

Municipality	*County*	*Area (square miles)*	*Population (2010)*
Shrewsbury Township	Monmouth	0.10	1,141
Loch Arbor Village	Monmouth	0.12	194
Lake Como Borough	Monmouth	0.25	1,759
Allenhurst Borough	Monmouth	0.25	496
West Wildwood Borough	Cape May	0.30	603
Cape May Point Borough	Cape May	0.33	291
Interlaken Borough	Monmouth	0.39	820
Ocean Gate Borough	Ocean	0.44	2,011
Avon-by-the-Sea Borough	Monmouth	0.46	1,901
Seaside Heights Borough	Ocean	0.51	2,887
Farmingdale Borough	Monmouth	0.53	1,329
Longport Borough	Atlantic	0.57	895
Englishtown Borough	Monmouth	0.58	1,847
Bradley Beach Borough	Monmouth	0.60	4,298
Allentown Borough	Monmouth	0.62	1,828
Pine Beach Borough	Ocean	0.64	2,127
Bay Head Borough	Ocean	0.70	968
Mantoloking Borough	Ocean	0.70	296
Lavalette Borough	Ocean	0.71	1,875
Highlands Borough	Monmouth	0.72	5,005
Neptune City Borough	Monmouth	0.88	4,869
Island Heights Borough	Ocean	0.89	1,673
Ship Bottom Borough	Ocean	0.99	1,156
Barnegat Light Borough	Ocean	1.00	574
Lakehurst Borough	Ocean	1.01	2,654

Source: Area: New Jersey Department of Environmental Protection, "NJ Geo-Web," http://www.state.nj.us/dep/gis/geowebsplash.htm (1996–2015). Population: U.S. Census Bureau, 2010 Census.

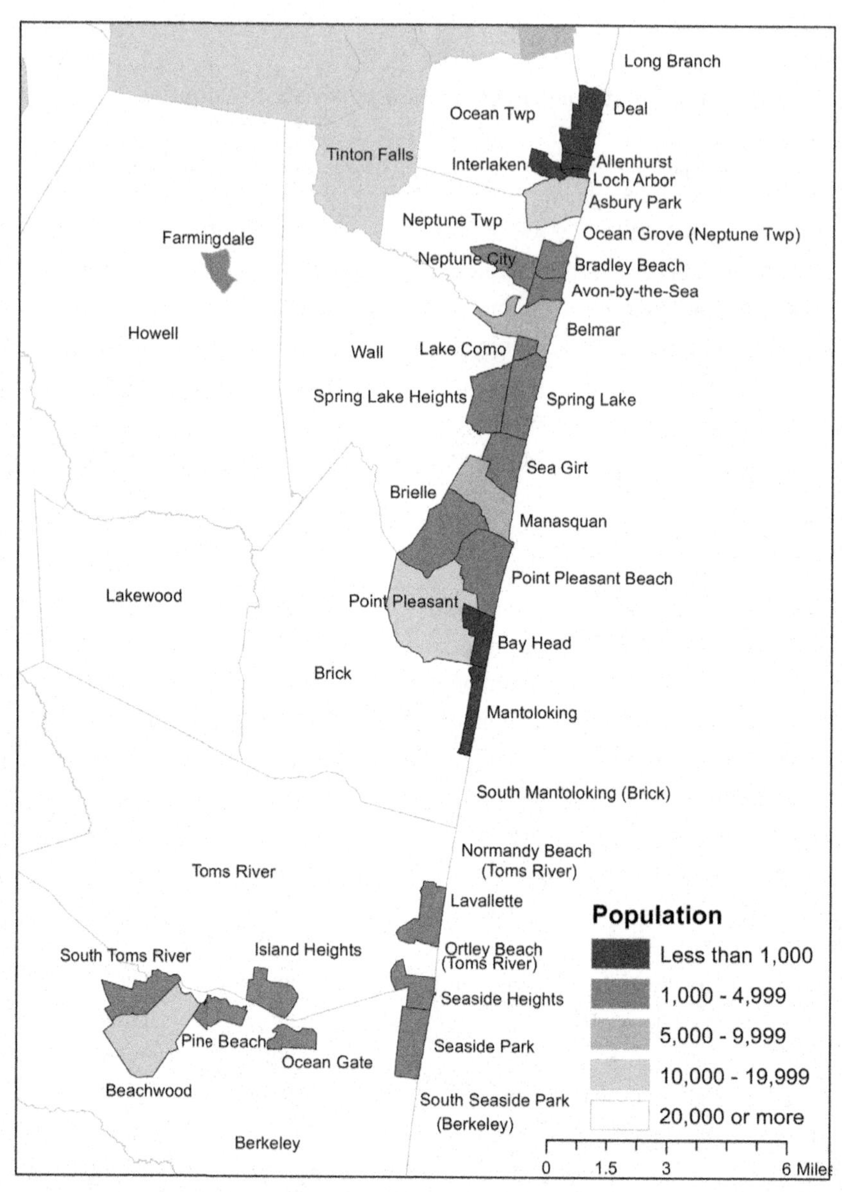

MAP 5. Micro-municipalities on southern Monmouth/northern Ocean Counties. Map by author, based on GIS shapefiles from New Jersey Department of Environmental Protection, "NJ-GeoWeb."

lead individuals to question the efficacy of government and feel as though the bureaucracy is working against them.

This situation is further complicated on the Jersey Shore because some oceanfront communities exist as culturally distinct places and may even be officially recognized as "Census-designated places" (CDPs), but they are administratively part of a larger municipal government.[9] This is true in Ocean Grove, which is part of Neptune Township, and Ocean County's three large oceanfront municipalities (Berkeley, Brick, and Toms River), where ocean-beach communities are physically separated from the mainland population centers by Barnegat Bay. Ortley Beach, Normandy Beach, and Ocean Beach, for example, are all part of Toms River; South Mantoloking is part of Brick, and Berkeley Shores and South Seaside Park are part of Berkeley Township. These beach communities have strong local identities and often have local quasi-governmental organizations, such as the Ocean Grove Homeowners Association,[10] the Normandy Beach Improvement Association,[11] the Ortley Beach Voters and Taxpayers Association,[12] or the Ocean Beach "clubs" that have homeowner covenants and deed restrictions.[13] While these organizations do not replace municipal administration, they provide locals with a stronger sense of belonging than "distant" municipal governments. These organizations allow participation by property owners who are nonresidents, an important improvement in representative governance in communities with large numbers of second-home owners. For example, in Ocean Grove (CDP), 29.5 percent of housing units are seasonally vacant; this number rises to 71.7 percent in Ortley Beach (Dover Beaches South CDP) and 81.8 percent in Dover Beaches North CDP, which includes Normandy Beach and Ocean Beach.[14] Local, quasi-governmental organizations also respond directly to the needs of the immediate community, while municipal governments must also respond to the larger number of year-round residents, who often live in neighborhoods unlike those on the barrier island.

Despite the potential for municipal hiccups given these convoluted local circumstances, few homeowners (year-round or seasonal) expressed disappointment about the actions of the local government after Sandy. In fact, locals expressed sentiments of solidarity within the community, which included the municipal officials, many of whom also suffered damage to their homes during the storm. Once it became clear that there was relatively little loss of life on the Jersey Shore, locals largely came together to help one another. Support did not just come from the neighbors who rode out the storm together, but from a variety of individuals, organizations, and agencies that rapidly converged on the region to begin the recovery process. To discourage and prevent looting, access to many beachfront communities was limited to residents and homeowners and enforced by local and state police as well as more than four thousand activated members of the New Jersey National Guard, so initial assistance took place mainly among people who shared losses and had close ties to the Shore. Within days, residents were joined by armies of volunteers offering resources, labor, and expertise. These experiences tended to strengthen community bonds and "faith in humanity," what sociologists more clinically refer to as "social solidarity". A Monmouth University/*Asbury Park Press* survey taken near the anniversary of the storm's landfall found that 81 percent of residents in the hardest-hit areas believed that the storm "brought out the best in people," compared to 74 percent of residents outside these areas.[15]

The Federal Presence at the Shore

Was this solidarity extended to representatives of higher levels of government, where face-to-face interactions were less common, as social dissimilarity and social structure reasserted themselves over time, and bureaucratic organization reigned? The short answer to this is: sort of. The Monmouth University/*Asbury Park Press* poll conducted a year after Sandy found that a greater proportion of people rated the "job done" by local government as "excellent" (27 percent) when compared to county (17 percent), state (22 percent), or federal government (17 percent), but only a minority rated any level of government as having done a fair or poor job (20 percent for local and county, 21 percent for state, and 27 percent for federal).[16] After the initial impact and immediate recovery were under way, perceptions of the behavior of government and other large actors began to become more antagonistic, then negative. Frustration and betrayal became common narratives as families attempted to navigate the process of rebuilding with the assistance of a variety of government agents, private insurance agents, and private contractors.

Antagonism toward the federal government came first, following the December 2012 release of preliminary Advisory Base Flood Elevation (ABFE) maps that indicated which areas of the Shore would require structural elevation to avoid steep increases in flood insurance (see map 6 for an example). The base flood elevation indicates where the Federal Emergency Management Agency (FEMA) estimates that ordinarily dry land stands at least a 1 percent chance of flooding on any given year.[17] Within this area, FEMA separates land into A Zones and V Zones, with the latter indicating locations where, in addition to floodwater, structures may be subject to currents or waves of three feet or more. The flood maps are the source material for Flood Insurance Rate Maps (or FIRMs), which are used to determine the cost of insuring structures against floods, through both private insurers and the FEMA-administered National Flood Insurance Program (NFIP). New flood maps had been under revision long before Sandy's landfall, but FEMA hastened the release of preliminary, advisory maps after the storm, in order to guide reconstruction without delay. The day after the release of these maps, however, a *Star-Ledger* headline warned "Jersey Shore Revolution Begins."[18] The maps nearly doubled the number of homes in the highest-risk V Zone, where homes would have to be elevated onto pilings. Signs from the grassroots social media–based StopFEMANow sprang up through the Shore region, though mainly in Ocean County, and local media began to report story after story of residents and property owners fighting large, unsympathetic bureaucracies.[19] FEMA did release revised maps in June 2013 that removed 45 and 46 percent of homes from the preliminary V Zone in Ocean and Monmouth Counties respectively. In Atlantic County 80 percent of homes were removed.[20] Nonetheless, the perception persists that FEMA maps will render the Jersey Shore uninhabitable except perhaps by the wealthy.

The flood map imbroglio is part of the larger problem associated with the increases in rates that homeowners will pay for the NFIP. As with FEMA's flood maps, flood insurance reform had occurred before Sandy struck. Since 1968, NFIP has provided subsidized insurance rates to homeowners who have agreed to follow federal maps for flood hazard mitigation. In the ten years prior to Sandy, however, the flood insurance program had greatly outstripped its own resources, mainly as a result of the 2004 hurricane season, when four major storms made landfall in Florida, as well as the mind-boggling devastation of Katrina the following year. A FEMA brochure released earlier in the month that Sandy made landfall explained that floods had caused more than $25 billion in damages since 2002.[21] National Flood Insurance Program reform came from the Biggert-Waters Act of 2012, which reauthorized NFIP through 2017, with the provision that it become financially sustainable.

Several elements in flood insurance reform were of great concern for the Jersey Shore. At its inception, the national program allowed flood-prone properties built before the initial flood maps were created to receive subsidized insurance rates. In New Jersey, initial flood maps were drafted throughout the 1970s, which meant that many Shore homes received federally subsidized flood insurance. For example, in flood-prone Sea Bright, more than 40 percent of housing units were built before 1970; in Long Beach Island's Beach Haven, this figure rose to nearly 58 percent.[22] Although later construction would take into account the flood maps and build at higher levels (or pay exorbitant flood insurance rates), the traditional Jersey Shore beach bungalow was generally buffered from these higher costs.

The Biggert-Waters Act removed these subsidies in 2012, with rates increasing 25 percent per year until the cost of the insurance reflected the true cost of likely flood damages ("full risk rate"). Biggert-Waters also required that full risk rates would be immediately applied if existing flood policies lapsed, if a new flood insurance policy was written, if the property was sold, or if repetitive claims were made on the same property. Biggert-Waters also ended subsidies

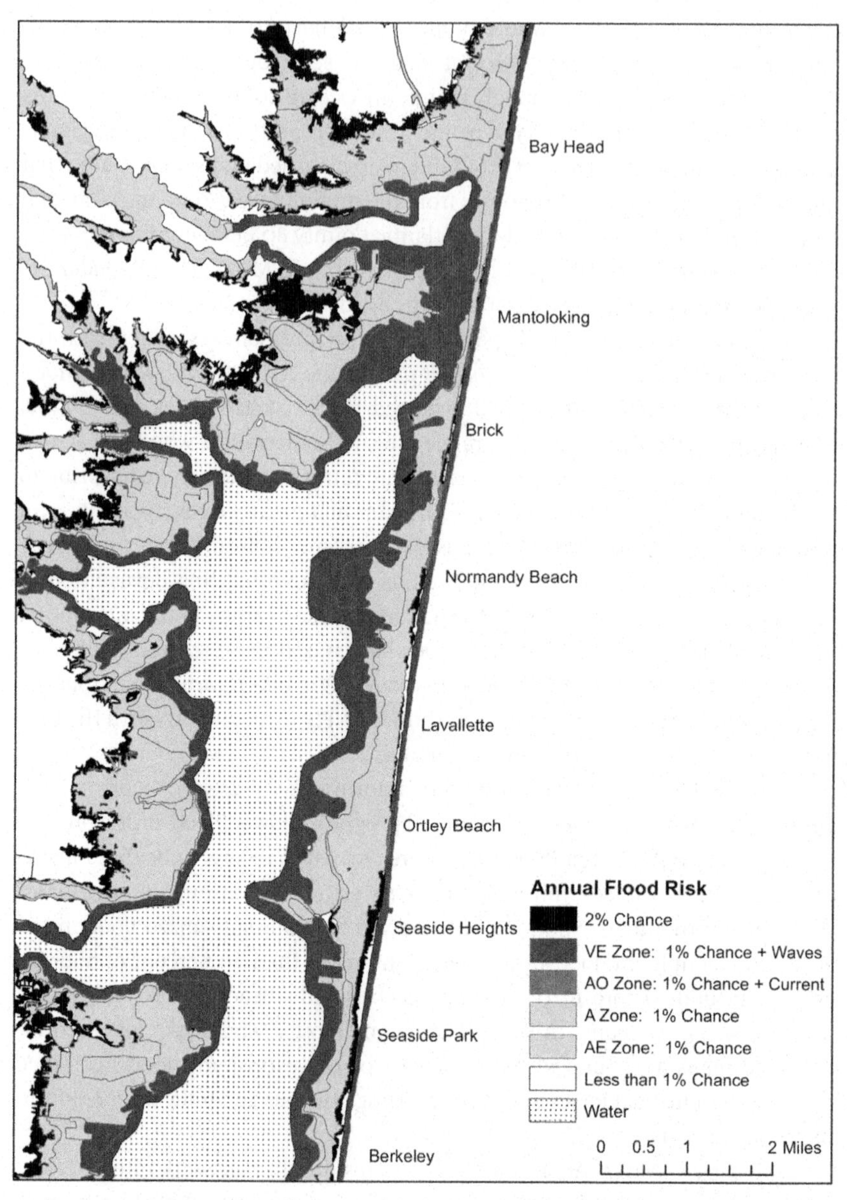

MAP 6. Flood insurance rate map, northern Ocean County coast. Map by author, based on GIS shapefiles from United States Federal Emergency Management Agency National Flood Hazard Layer Map Inventory, http://floodmaps.fema.gov/NFHL/status.shtml.

for properties that had made claims for greater than the market value of homes, and for properties with repeated severe claims. Finally, Biggert-Waters eliminated or phased out subsidies for both second homes and businesses, which were expected to convert to full risk rate policies before 2017. All of these provisions would undoubtedly increase costs for Shore residents, and for those who were now facing the additional costs of restoration from Sandy, they were unbearable.

In response, New Jersey Senator Robert Menendez wrote what would become the Homeowner Flood Insurance Affordability Act of [March] 2014, which moderated some of the provisions of Biggert-Waters. Modifications included a cap on rate increases of 18 percent annually (although in some cases, this could be limited to 5 percent annually). The Homeowner Flood Insurance Affordability Act also eliminated the immediate increase to full risk rate associated with property sales and the issuance of new flood insurance policies—critical in the aftermath of Sandy. It also reinstated "grandfathering," so that existing homes that had their risk increased because of new maps would have a more gradual increase in insurance rates, although increases would be a minimum of 5 percent and a maximum of 18 percent each year until full risk rate was reached. Substantively, the 2014 law did not change Biggert-Waters provisions regarding older businesses and continued to exclude second homes from receiving subsidized rates, nor did it protect properties that had sustained repetitive damages. As a consequence, many homeowners rebuilding after Sandy decided to sell or to rebuild differently. Jersey Shore neighborhoods are now full of smaller and older homes undergoing elevation and demolition, destroying the traditional bungalow landscape of Shore neighborhoods. Oceanfront bungalows are even less likely to survive; the *New York Times* reported the evaluation of one Shore realtor: "Most buyers are tearing down what exists and building large elevated houses according to the stricter guidelines and adding elevators, swimming pools, and walls of windows."[23]

The State Government: Et Tu, Trenton?

The state government, and especially Governor Chris Christie, enjoyed remarkably strong bipartisan support immediately after the storm. A November 2012 Rutgers-Eagleton poll found that Christie enjoyed a 67 percent approval rating, "one of the highest approval ratings ever enjoyed by a New Jersey governor."[24] Christie's popularity surged following his January 2013 rant against the "toxic politics" and "duplicity" of the Republican national leadership, which had stalled Sandy relief aid as part of its campaign to reduce federal budgets.[25] Christie easily won his November 2013 reelection, which prominently featured the governor's activities during and after Sandy.[26]

Despite the continued popularity of Governor Chris Christie, goodwill for state government began to erode as the rebuilding process continued. In January 2013 the state decided to adopt FEMA's controversial preliminary flood maps, which state officials justified as facilitating reconstruction by providing noncontradictory guidelines.[27] Confidence in the governor himself waned in the wake of the George Washington Bridge scandal in September 2013, and subsequent allegations from Hoboken mayor Dawn Zimmer that a member of the Christie administration threatened to withhold Sandy aid until she approved a real estate development that the governor favored.[28] Public approval of the governor and state government dropped again following the release of a report indicating that hard-hit Ocean County received only 7.7 percent of the initial disbursement of $160 million of federal Housing and Urban Development Community Development Block Grant (HUD CDBG) money earmarked for affordable housing for Sandy victims in New Jersey, while projects far from storm-affected areas were approved.[29]

In an April 2014 survey, a Monmouth University/*Asbury Park Press* Poll found that fewer than half (48 percent) of residents in the hardest-hit areas thought that the state was doing a "good job" with Sandy relief efforts, while a nearly identical proportion (47 percent) thought the state was doing a "bad job."[30] Residents in these areas attributed delays in assistance evenly to both federal regulation (43 percent) and management by the state (42 percent).[31] The state's reNew Jersey Stronger's Reconstruction, Rehabilitation, Elevation, and Mitigation program (RREM), managed by the Department of Community Affairs (DCA), promised to channel $600 million from federal funds into the region. A panel study of residents affected by Sandy conducted by the Monmouth University/*Asbury Park Press* Poll found that 83 percent of those who applied received a $10,000 RREM resettlement grant if they relocated within the county, but only 9 percent successfully secured grants for reconstruction or rehabilitation of up to $150,000 and only 15 percent received grants for elevation of up to $30,000.[32] Not surprisingly, nearly half of respondents felt they were denied assistance that they deserved. In the same vein, a Rutgers-Eagleton poll conducted statewide also found that by April 2014, 31 percent of respondents who had personally been affected by Sandy indicated that the New Jersey state government had handled recovery efforts "somewhat" or "very" badly, and the governor's approval rating had dropped to 50 percent.[33]

The conflation of the administrator—such as Governor Christie—with the government's bureaucratic apparatus is just one problem with trying to analyze how and why New Jersey residents increasingly feel that the state and the federal government have violated their contract with the public. Partisanship is rampant—81 percent of people who approved of Christie's performance as governor also indicate that the state is handling Sandy recovery "somewhat" or "very" well, while only 49 percent of those who view the governor unfavorably concur. Christie has encouraged the conflation, assigning Sandy coordination efforts in the state government to the "Governor's Office of Recovery and Rebuilding."[34] The governor's wife, Mary Pat Christie, was appointed to chair the high-profile nonprofit Hurricane Sandy New Jersey Relief Fund, which has collected over $40 million in donations.[35] Of the sixty press releases issued by the New Jersey Office of Emergency Management (OEM) since Sandy, only twenty-four do not include "Christie" in their headline, and in fourteen of these the governor is the actor in the headline—often gratuitously, as in "Governor Christie Releases Power Utility Companies' Revised Plans for Service Restoration."[36] It should be noted, however, that although the governor enjoys executive powers within the state bureaucracy, much of the public policy that determines how government interacts with the environment predates the current administration and will continue to shape interactions with the Jersey Shore long after Chris Christie leaves office.

Although many do not believe it, bureaucratic organization exists to rationalize the complex operations of modern society. Bureaucracies rarely live up to this ideal, however. Modern bureaucracies are typically slow moving and unresponsive, and are criticized for elevating rules ("red tape") over outcomes. In the state government, three bureaucracies are particularly important for understanding Sandy's recovery efforts: the Department of Environmental Protection, the Department of Community Affairs, and the Office of Emergency Management.

Emergency response officials, particularly first responders, typically enjoy the highest level of public support.[37] The public understands emergency response, and firefighters, police officers, and emergency medical technicians (EMTs) are widely recognized for their selfless professional behavior. Behind the scenes, planning for emergencies is the main function of the New Jersey Office of Emergency Management (OEM), which coordinates county and municipal emergency management offices, and acts as a liaison between local and federal agencies, including FEMA. When the federal government expanded its national civic defense program to include an "all-hazards" approach, New Jersey created a statewide OEM.[38] As with FEMA's location within the Department of Homeland Security, the OEM's activities are administered by the Homeland Security Branch of the New Jersey State Police, whose superintendent has been historically cross-appointed as the director of the New Jersey Office of Emergency Management.[39]

The New Jersey OEM is "responsible for planning, directing and coordinating emergency operations within the state that are beyond local control."[40] It is organized into three bureaus: Communication, Emergency Response, and Recovery. The Communications Bureau handles incoming emergency calls (such as 911 calls) and coordinates dispatch of first responders and other communication.

5.1. State police patrol Seaside Heights boardwalk, winter 2012. Photo by author.

The Emergency Response Bureau coordinates the first responders to emergencies, including hazardous materials cleanups and search-and-rescue operations.

The Recovery Bureau consists of five units: Public Assistance, Preparedness, Mitigation, Field Training, and Support Services. The Public Assistance Unit surveys officially declared disasters to determine what is needed, and it helps coordinate and provide direct aid in the immediate aftermath. The Preparedness Unit helps guide and approve disaster preparation plans, as well as conducting public outreach and education. The Mitigation Unit provides information and support for municipalities seeking grants to reduce risk of recurring disasters, mainly from federal programs such as the Flood Mitigation Assistance (FMA), Pre-Disaster Mitigation, and the Hazard Mitigation Grant Program. The Field Training Unit coordinates emergency training exercises. The Support Services Unit coordinates volunteer programs. These programs are critical to disaster planning and preparation, but funding for all state-funded law enforcement activities in New Jersey was reduced by just over a million dollars from 2013 to 2014. The entire Office of Homeland Security and Preparedness, which includes the Office of Emergency Management, had a budget in 2014 of just $3.7 million.[41]

Beyond an appreciation of first responders and emergency management, lay understanding of government bureaucracies and natural disasters get increasingly muddled. People understand that the government regulates environmental impact, but few understand the intricacies of how this actually occurs, or what the role of an environmental agency would be in a disaster like Sandy.[42] The New Jersey Department of Environmental Protection (DEP) is the agency charged with protecting the state's environment. The DEP officially came into existence on the first Earth Day, April 22, 1970, nearly eight months before the federal Environmental Protection Agency (EPA) was created. The same year, the DEP mapped the salt marshes that would be protected under the state's Wetlands Protection Act of 1970.[43] This law was followed by the Coastal Area Facilities Review Act (CAFRA) of 1972, which legally defined New Jersey's "coastal areas."[44] In addition to delineating endangered ecosystems, these laws created an obligation for the state to determine threats to the region's environment, set standards for construction and reconstruction, and enforce those standards. Although both the Wetlands Protection Act and CAFRA have some ecocentric passages, neither law was designed to halt or reverse coastal development, but rather to reduce the environmental impact of existing and future construction. These laws explicitly took into consideration the economic and social importance of coastal development; from the outset, the regulation of human development on the Jersey Shore was based on multiple interests and stakeholders. Considerations in the laws included threats to nonhuman species, as well as the potential dangers for people who live along the Shore. For example, the Wetlands Act of 1970 included the following passage: "In granting, denying, or limiting any permit, the commissioner shall consider the effect of the proposed work with reference to the public health and welfare, marine fisheries, shell fisheries, wildlife, the protection of life property from flood, hurricane, and other natural disasters."[45] As CAFRA further regulated coastal development, the focus on multiple stakeholders persisted. CAFRA and the state's consolidated Coastal Management Rules, which brought the state into compliance with the Federal Coastal Zone Management Act of 1972, are multifaceted, running to eight broad goals and thirty-one subgoals. Coastal Management Rules mention natural hazards in two goals ("Coastal open space" and "Safe, healthy, and well-planned coastal communities"), but these appear in subgoals and follow larger concerns about ecosystem health, natural resource management, public access, and economic development and revitalization.[46]

The administrative guidelines set forth by the state's Department of Environmental Protection reflect how long-term environmental and land-use planning affect and are affected by disasters such as Sandy. However, planning for natural disasters is generally treated as a secondary or tertiary concern for the existing laws that govern land-use planning in New Jersey's Coastal Zone. DEP-enforced laws prioritize ecosystem health and economic impact over hazard mitigation. Like the OEM, the DEP also operates on a shrinking budget: despite Sandy's impact, the DEP's entire budget in 2014 was smaller than its 2013 budget by more than $13 million, amounting to $328.2 million.[47]

The third state bureaucracy relevant to disaster response in New Jersey is the Department of Community Affairs (DCA), which, according to its own website, was created to "provide administrative guidance, financial support, and technical assistance to local governments, community development organizations, businesses, and individuals to improve the quality of life."[48] In other words, the DCA is involved in all sorts of oversight and promotion of public, private, and nonprofit interests, including divisions in building codes, fire codes, housing, and government oversight. The DCA also created a new Sandy Recovery Division and authored the official "Action Plan" for associated recovery programs.[49] This document outlines the agency's responsibility to conduct an assessment of "unmet needs" in recovery efforts and to design, implement, and assess the programs intended to meet those unmet needs. The disbursement and tracking of the $1,829.5 billion in grants awarded as part of the federal "Sandy Supplement" was controlled by the DCA, although the state comptroller tracks and publishes this and other public spending through its NJ Sandy Transparency program.[50] Federal Housing and Urban Development (HUD) Community Development Block Grant (CDBG) funds require attention to both the areas that suffered the most damage and to low- and moderate-income populations. The DCA's own unmet-needs assessment indicated that while Ocean and Monmouth Counties suffered the greatest losses in the state, the highest proportion of low- and moderate-income residents severely affected by the storm was in North and Central Jersey, well away from the Jersey Shore. For example, 71.3 percent of households with "major/severe damage" from Sandy in Essex County were low- and moderate-income households, mainly in Newark along the flooding Passaic River, compared to only 45.2 percent in Ocean County.[51] The mission and constraint of federal grants helps explain why Ocean County has received proportionately less grant money, even though this county was arguably the worst affected by the storm.

Recreancy and Betrayal

One of the most influential theorists in environmental sociology, William Freudenberg, found that what people thought the government should do and what the government was actually able to do were often at odds with one another. Moreover, governmental agencies were often limited by regulatory and budgetary boundaries; as a consequence, government agencies may be unable to do what people want them to do before, during, and after disasters. Freudenberg called this "recreancy," which he defined as "a retrogression or failure to follow through on a duty or trust . . . [or] behaviors of persons and/or of institutions that hold positions of trust, agency, responsibility, or fiduciary or other forms of broadly expected obligations to the collectivity, but that behave in a manner that fails to fulfill the obligations or merit the trust."[52] Later, Valerie Gunter and Steve Kroll-Smith distinguished two important types of recreancy: "premeditated betrayal," in which an individual, agency, or organization deliberately violates its responsibilities, and "structural betrayal," in which an agency or organization appears to violate a social contract but are in fact institutionally limited.[53] Gunter and Kroll-Smith also acknowledge that in most cases, it is impossible to determine if "failure to follow through on a duty or trust" was a deliberate, premeditated action or the result of institutional constraints. As a consequence, many failures are "equivocal," with no clear suggestion of why institutions failed, or are perceived by the public to have failed.

Recreancy, or perceived betrayal by government officials, is thus often explained by what Gunter and Kroll-Smith call "structural betrayal." For example, the OEM is hampered by its role as a coordinating organization. The DEP has mission-based priorities that supersede storm hazard mitigation. Both the OEM and DEP have limited and declining budgets. The DCA, flush with federal Sandy relief money, is constrained by the strings attached to that money. The state's affected residents fault the government for impersonal treatment and apparent lack of compassion, but these are in part the result of limits to bureaucratic authority and the noisy political demands for financial accountability in the disbursement of public funds.

The public grows wary and suspicious of recreant behavior when actual corruption is revealed among public actors. This happened in New Jersey with the George Washington Bridge scandal, in which the governor's deputy chief of staff and two major political appointees resigned amid mounting evidence of the political motivation behind lane closures on the bridge in September 2013, and the subsequent cover-up of the decision to create the traffic jam. Although wholly unrelated to the Sandy recovery effort, investigations into "Bridge-gate" spurred Hoboken mayor Dawn Zimmer to come forward with additional allegations of improper political maneuvering by members of the Christie administration. Although the events were still under investigation at the time of this writing, Zimmer claims that Lieutenant Governor Kim Guadagno threatened to withhold Sandy relief money if Zimmer did not approve development permits desired by Christie supporters. Zimmer's accusations are used by Christie detractors to suggest that Sandy aid is being distributed in an inappropriate and perhaps corrupt way, thus violating the social contract of the state government with the victims of the storm. Until or unless Zimmer's claims can be substantiated, however, this is a better example of "equivocal betrayal," in which it may appear that the government has breached faith with the public and many people will believe that it did. Gunter and Kroll-Smith warn that "once individuals believe a conspiracy is afoot, they come to interpret all kinds of actions, inactions, and utterances through that framework."[54] As a consequence, even public agencies and agents acting in good faith (but within structural limits) can be interpreted as recreant by those who are inclined to believe in widespread corruption in the Christie administration.

Technical Fixes to Social Problems

People's attitudes about the government are also tempered by a widespread belief that social problems—especially environmental problems—have technical solutions. Rather than questioning the fundamental, underlying patterns of social organization, people increasingly turn to engineers for solutions. Public policy reflects this preference; consequently, government responses to Sandy have been predominantly technical. For example, rather than questioning the basic decision to build permanent structures on ephemeral landforms like barrier islands, governments at all levels have proposed new building codes, including elevation of structures.

Science and engineering do not always provide the definitive answers that people desire, however. For example, Mantoloking and Bay Head are pursuing different technical strategies for preventing future flooding: protective dunes and sea walls. Bay Head has a protective sea wall and rock groin, initially built in 1882 and designed to stabilize a line of dunes. An engineering survey made after Sandy reports that the sea wall was critical in reducing the damage caused by wave action during the storm.[55] Mantoloking contends that the sea wall intensified the flooding and damage to their town by pushing the surge away from Bay Head, and extending the sea wall will make things worse for Mantoloking in future storms.[56] Instead of building a sea wall of their own, Mantoloking, like much of Long Beach Island, has instead invested in beach replenishment and increasing the size and stability of the protective dunes.

Mantoloking's plan is consistent with the $158 million plan of the U.S. Army Corps of Engineers and New Jersey DEP for the shoreline between the Manasquan and Barnegat Inlets, which calls for a higher dune berm and wider beaches, but no sea wall.[57] Which engineers have the better solution? The state and federal government has clearly sided with the Army Corps of Engineers; Governor Christie infamously called oceanfront homeowners who refused to grant easements for dunes "knuckleheads," said that their position was "bullshit," and warned: "We are building these dunes whether you consent or not."[58] The state has backed these statements with a $65 million commitment to beach and dune replenishment in Monmouth County,[59] as well more than $23 million in Ocean County.[60]

What is curious about the decision to spend public money on dune berms and beach replenishment is that even the engineers acknowledge that dune construction is far from a permanent solution. For example, the Army Corps of Engineers' long-term plan for Long Beach Island builds in substantial replenishment

5.2. Established dune, Island Beach State Park, summer 2013. Photo by author.

5.3. Dune grass planted in front of doomed Funtown Pier, Seaside Park, summer 2013. Photo by author.

every seven years.[61] This, like other Corps projects along the Jersey Shore, will require substantial additional injections of money following destructive storms, of which Sandy is only one of many. Rarely discussed is the DEP-administered program (largely funded by FEMA) known as Blue Acres Floodplain Acquisition, which offers voluntary buyouts for properties that have suffered repetitive flood losses. Following Sandy, all Blue Acres buyouts have been in the Raritan River/Bay region, purchasing forty-four properties in South River and Sayreville. The DEP has funds for purchase of an additional 380 properties, all of which are in the Raritan River/Bay area, along the Passaic River in Newark, or along the Delaware Bay.[62] Some environmental organizations have criticized the lack of discussion about whether waterfront properties *should* be rebuilt after Sandy, but this has failed to translate into public policy. An October 2013 press release from the Sierra Club New Jersey chapter states unequivocally, "There is no planning for pulling back from environmentally sensitive areas," and Jeff Tittel, the chapter director, is quoted as saying, "We seem to be doing some of the same mistakes over again when it comes to rebuilding our coast and protecting our state from floods and storms."[63]

Relying on technical tweaks without altering the larger social systems is consistent with a theory in environmental sociology called ecological modernization. Ecological modernization primarily focuses on changes to industrial production to reduce environmental impacts. The basic assumption of ecological modernization is that as people become more attuned to environmental concerns, especially pollution or depletion, they will demand modifications in the social world to conserve and protect the environment. Conservation and protection can be done without fundamentally disrupting the basic structures of society and can be achieved mainly through improvements in technology. The classic example of ecological modernization is the adoption of the catalytic converter, which has reduced air pollution from individual cars. Regulation can be a stimulus for technological change: the technology was introduced widely into U.S. automobiles as a response to both the Clean Air Act of 1970 and federal clean fuel standards beginning in 1973, which reduced and eventually eliminated lead additives to gasoline.

Ecological modernization is also relevant to understanding how New Jersey prepared for and responded to Sandy. Construction engineering has more or less stabilized the oceanfront for permanent occupation and infrastructure development, even on the shifting sands of barrier landforms. Land-use regulation, including flood maps and CAFRA, should indicate structures at greatest risk. When threats do appear, emergency management technologies (for example, weather services, evacuation, and search-and-rescue technologies) have reduced the likelihood of loss of life and property.

Proof of technical preparedness came just a year before Sandy's arrival. On August 28, 2011, Hurricane Irene made landfall in Brigantine. Irene caused a three- to five-foot storm surge along the Jersey Shore and inundated the state's northern rivers with what the National Weather Service described as "torrential" and "record breaking" rains.[64] Atlantic City closed its casinos for only the third time since 1978. A 2012 DCA report indicates that more than one thousand New Jersey homes were "severely damaged" by the storm.[65] Mandatory evacuations along the Jersey Shore were executed more or less successfully, and emergency management officials threatened recalcitrant residents with dire examples from Hurricane Katrina: Cape May County Emergency Management Director Frank McCall was reported to have told those who ignored mandatory evacuation orders to put their social security number and next of kin on an index card in their left shoe, so their bodies could be identified in the recovery process.[66] Although seven New Jersey residents died as a direct result of Irene and tens of thousands went without electricity for a week or more, the National Weather Service concluded that forecasting and communicating risk of the storm were considered among "the most successful [National Weather Service] activities during Irene. NWS staff embedded in emergency operations centers were *universally praised* for their consistent, authoritative, and neutral message" (italics in original), and the Philadelphia/Mount Holly Forecast Office was singled out for producing "high quality and timely" reports for the local emergency management services.[67] Moreover, repairs made after Irene were designed to "harden" forecasting infrastructure against future storm damage.

A year later, public actors and emergency management officials were once again able to effectively convey the risk of the pending storm. In its "Service Assessment of Sandy," the National Weather Service reported that surveys with emergency management professionals, the media, and the general public indicated that weather and damage forecasting were accurate and useful. Evacuations in coastal areas went well, although one notable exception to this was Atlantic City. Mayor Lorenzo Langford feared that the largely unnecessary evacuation of Absecon Island for Irene the previous year would lead many residents to remain in Atlantic City during Sandy. While strongly encouraging residents to evacuate, Langford authorized the opening of schools as shelters of last resort, earning the very public ire of Governor Chris Christie.[68]

After the storm, emergency assistance was once again well organized and rapid, with local first responders moving into affected areas before daybreak on October 30. These were followed by the timely deployment of the New Jersey National Guard to assist in recovery and maintain order. Sandy had unleashed unprecedented damage on the Garden State, but given the potential for catastrophic destruction, especially in comparison to Hurricane Katrina, preparations and response to the storm had largely worked well. With this outcome, few have questioned whether the Shore should be put back together the way it was.

The next chapter examines why people are not interested in having discussions about changing the Jersey Shore in the wake of Sandy. New Jersey residents—even those who suffered substantial property damage—do not deny the inevitability of future storms. Many waterfront residents fatalistically accept that periodic property losses are worth the benefit of living so near this natural amenity. The acceptance of the status quo is not always as simple as a cost-benefit analysis; humans are creatures of habit, and habits provide much-needed feelings of security. Restoration of the Shore may be important symbolically and economically, but the desire to "return to normal" is a siren song that drowns out the catastrophic risks inherent in dense coastal development.

6

Restoring Security at the Shore

Most of what organizes the modern social world is essentially invisible to us as we go through our daily lives. In other words, we do not spend most of our time thinking about the social organization that makes modern living possible. For example, this morning, I got up, took a shower, made some coffee, ate some yogurt and a banana, brushed my teeth, kissed my kids and husband goodbye, checked my e-mail, and then sat down to write. What was invisible to me? At the most obvious level, I expected there to be both electricity and running water. Without electricity, my alarm would not have woken me, I could not have turned on the light in the bathroom, I could not have made coffee nor stored my yogurt, and I certainly could not check my e-mail or write (since I write on a computer). Without running water, I could not have showered, made coffee, or brushed my teeth. At a more complex level, I depend on global food sources, with my yogurt coming from New York State, bananas coming from Guatemala, and coffee from Colombia. For my breakfast, I not only depend on the producers in these far-flung places, but all the intermediary shippers, wholesalers, and retailers (among others). I also rely on external institutions to organize my day; for example, my daughter goes to a public school, and my son to a private day care center. It is not really possible for me to work unless someone is looking after my kids, so I am dependent on these institutions to free up my time to work. I am also dependent on the transportation that allows my husband to drive my son to day care and himself to work, and the school bus that picks up my daughter, and these involve a whole separate global network of vehicle and fuel production. Transportation also requires maintenance of roadways and management of other drivers, which is done both by human enforcement of laws, and through electricity-dependent traffic lights. These tell drivers when to stop, go, and slow down and prepare to stop.

These invisible processes get even deeper when I begin to consider that I assume that all of these systems function the way they are "supposed" to and that they are not having unintended consequences for me or my family. I assume that the water that comes out of the pipes in my house is safe to drink. I assume that the food that I eat is safe to eat. I assume that the air I breathe is clean. I assume that the places where my family goes during the day are safe and secure. I assume that the people who are charged with taking care of roadways and global trade are doing their jobs. I assume that if I woke to find that there were no bananas in my house, I could go to the grocery store and buy more, and that I could do so using a piece of plastic (which is, in itself, an assumption of a functioning financial system, which assumes a stable government and economic institutions). I assume that there will be a grocery store, and that there will be people working there who can help me. I assume that there will be ample fuel available for me to get to the grocery store (which my magic plastic card will also provide). I assume that I will be able to get my fuel and groceries without having to fight through a zombie apocalypse. You can see at this point that there are a lot of assumptions I am making, even before lunch. And perhaps you've heard of what happens when you assume.

Modern life, with its complex, global division of labor and dependence on technology, is a scary place to be if you think too much about it. Consequently, we moderns generally *don't* think too much about it, enjoying what sociologist Anthony Giddens calls "ontological security."[1] Ontological security rests on the fact that most of the time, the systems on which we depend will function as we expect them to function. The day-to-day operations of these systems allows me to develop a psychological trust that those systems will continue to function. The more reliable these complex systems are, the more I can trust them on a day-to-day basis and allow them to fade into the background. I can live my privileged life as an academic in the New Jersey suburbs and not worry too much about how the world works. Unfortunately, there are times bad when things happen, and that ontological security is disrupted. Sandy was one of those times.

In October 2012, because of the weather forecasting, which I followed mainly through my cable television, I knew that Sandy was headed in my direction. I spent the weekend of October 27 and 28 making sure that we had candles, batteries, nonperishable foodstuffs, and drinking water. I also froze large bags of water, as I had for Irene a year earlier, as both a supply of water and a means to keeping the freezer cold, should we lose power. My father and his wife, who were visiting from out of town, had taken the train to my brother's house in Washington, DC; I called them and warned them that they would not be able to get back to New Jersey as originally planned, because all train service had been cancelled. We had been notified by our day care provider that the facility would not be open on Monday, October 29, as the governor had declared a state of emergency in preparation for the storm, but this was not a problem for me, because my college and my husband's workplace had also closed for the same reason. Prepared, my family and I spent Sunday literally at the park, taking some pictures at the nearby Delaware River and in our neighborhood park, as the outer bands of Sandy became apparent in our skies.

We woke on October 29 with the wind howling like a freight train outside—and we live about forty-five miles inland from the coast. The wind and rain continued all day long, although the rainfall was not particularly heavy. Cautious about leaving the house because of the high winds, which, in the suburbs, are scary because of what they can do to trees and power lines, my family left the TV on all day to keep up with the storm and checked in frequently with friends and family via phone calls and social media. Like people nationwide, we watched as part of the Atlantic City boardwalk washed away (although much less than was originally feared) and weather forecasters braved iconic positions along the Jersey Shore and New York City waterfronts to bring us updates. The storm grew stronger as the day went on, although as late as 7 PM, I was still posting snarky comments on Facebook. At a quarter to nine, weather radar put what would have been the eye wall over my house, so we felt somewhat confident that this was not going to be the big one after all. Not long after, we started to hear and then see the eerie green transformer explosions in our neighborhood, which drove my daughter to leave her bed and join us in front of the TV, although my then-one-year-old on slept through the whole thing. Then the power went out.

We lit candles, found our flashlights, cranked up the weather radio, and used our phones to update our Facebook pages to let people know that we were okay but had no power. My daughter eventually fell asleep, but my husband and I stayed up late listening to radio reports of the storm, pausing every now and then to crank the radio. My husband woke up before me the next day and took a walk through the neighborhood, finding lots of downed trees and power lines. We lost four trees in our own yard; the trees knocked down our fence but spared our house and cars. The utility crews quickly closed off one of the roads into our neighborhood, where a live wire lay across the road, but we spent most of the day doing storm cleanup and walking around talking to neighbors to assess damage. A few houses and cars in our neighborhood had trees fall on them or through them, and no one had power, but for the most part we did okay. As we learned more and more news from the radio, from talking to neighbors, and from our limited access to the Internet (through our phones, which we were charging from our cars), it became clear that other parts of the region had not been so lucky.

Friends and family members from Brooklyn to Margate were posting and reposting their own stories and pictures of damage along the Shore and in the New York area. Almost all of Hoboken was flooded and being evacuated. Lower

6.1. Tree on house, Ewing, October 30, 2012. Photo by author.

Manhattan was flooded and without power. Swaths of the Raritan Bay shore and the south shore of Staten Island were simply gone, including dozens of individuals who perished in the surge. Parts of Ocean County's barrier peninsula were burning, as gas mains had broken and eventually ignited. Photos emerged of places where the ocean had cut through the Barnegat Peninsula and barrier islands of the Jersey Shore. Millions of people were without power, but by the end of the day on October 30, we were sure that no one in our immediate circles was missing or seriously injured, although one friend's family had to be evacuated by helicopter and several fretted openly about the condition of their homes and second homes along the Shore. We went to sleep on October 30 after a candlelight dinner with our kids, bundled up against the rapidly dropping temperatures (we have forced air heating, so without power, we had no heat), and still listening to our hand-crank radio.

The next day was Halloween, which the governor rescheduled. My family became energy refugees and relocated to my in-laws' home in inland Toms River rather than face thirty-two-degree overnight weather with no heat or electricity. The drive to Toms River was a microcosm of recovery efforts. Interstate 195 was full of convoys of out-of-state utility vehicles heading east, but it was otherwise fairly free of traffic, since much of the state continued to be shut down. Off the freeway, roads were blocked with fallen trees and power lines, most traffic signals were without power, gas stations had either long lines or, more frequently, hand-painted signs reading "NO GAS" or "CASH ONLY." Few other businesses were open. We were joined in my in-laws' townhome by my husband's niece, her fiancé, and their dog, who had evacuated Hoboken and were unlikely to return to their jobs in Lower Manhattan any time soon.

Although it had suffered no major damage, my college was closed for the remainder of the week to allow students, faculty, and staff to attend to recovery efforts. I spent my days commuting home to clean spoiled food out our refrigerator and work on the damage to our trees and fence. Because I live close to Pennsylvania, where there was no shortage, I never had to wait more than fifteen minutes to get gas, but each afternoon when I returned to Toms River, fewer and fewer gas stations seemed to have fuel. This was not the only clue that things were far from back to normal. One evening, my husband and I traveled west down State Highway 37, which ordinarily travels across the Toms River Bridge to Seaside Heights. The National Guard had set up at the base of the bridge, and the traffic signs blinked ominously that the bridge was closed. We stopped for a drink at a nearby watering hole, only to find that it had no heat and could only accept cash, but we were welcome to stand about in our winter jackets with the other people who were tired of being cooped up for a week, many of us living out of suitcases in other people's homes. On Saturday, November 3, I was working in the yard when I spotted two utility trucks on my block. Later that day, our power was briefly restored, then went out again, but was completely restored by evening. We took our kids home that night and life more or less returned to normal. In all, my routines were lightly disrupted for about a week, which was not enough to shake the psychological trust that I had built up over the years preceding Sandy. But many were not so lucky.

Damage and Disruption to Daily Life

The United States Agency for Housing and Urban Development (HUD) defines "major damage" from a storm as when a home needs repairs costing between $8,000 and $28,799 and "severe damage" as costing more than $28,800.[2] By January 2013, the New Jersey Department of Community Affairs (DCA) estimated that 56,077 homes in New Jersey had suffered major or severe damage as a result of Sandy, of which about 40,500 were primary residences.[3] The vast majority (98 percent) of these were contained in nine counties, and more than 80 percent of the total were in the four coastal counties (table 6.1). Some communities were absolutely devastated. The DCA identified twenty census tracts on the Shore where more than 50 percent of housing units had major or severe damage; of these, thirteen tracts had more than two-thirds of units so damaged, and in four (in Middletown, Little Egg Harbor, Brick, and Toms River), more than 90 percent of homes had major or severe damage. By November 2, most of the Barnegat Peninsula, from Mantoloking to Island Beach, was under mandatory

TABLE 6.1.

Major and Severe Damage to Housing Units from Sandy by County

	Percentage of units with "major" or "severe" damage	*Number of units with "major" or "severe" damage*
Coastal Counties	*9.3*	*44,897*
Atlantic	9.0	8,744
Cape May	5.0	2,446
Monmouth	5.0	11,467
Ocean	10.3	22,240
Other Counties	*0.8*	*8,270*
Bergen	1.0	2,848
Essex	0.1	397
Hudson	1.9	4,407
Middlesex	0.7	1,975
Union	0.4	643

Source: New Jersey Department of Community Affairs, "Community Development Block Grant Disaster Recovery Action Plan," 2013, Section 2 (pages 2-3–2-11).

evacuation and had its utility services shut down, as emergency crews went in to survey and repair damage.

Many damaged communities remained "depopulated" in the weeks and months that followed. The DCA estimated that at least 44,000 displaced households received rental assistance after the storm, and these figures did not include the countless numbers who did not apply for aid, or the owners of second homes. Long Beach Island (LBI) homeowners were allowed to return to the damaged island within two weeks of the storm, but municipalities on the Barnegat Peninsula allowed residents only brief visits. Seaside Heights organized a bus that allowed just two residents with two suitcases each per property for their first post-Sandy visit home in early November.[4] Later, as officials became more confident in the infrastructural repairs, private vehicles were allowed, provided that occupants could demonstrate property ownership at checkpoints staffed by local police, state police, and the National Guard. Curfews remained in effect throughout the winter, with both police and armed National Guardsmen patrolling. By the end of December, businesses in some locations, including Seaside Heights, were open for nonresidents who were curious to see the state of the Shore, but all had to leave the area by the 4 PM curfew. There were also police roadblocks outside the commercial areas, to discourage looting. Mantoloking was the last town to allow residents back, in late February 2013, nearly four months after the storm.

Obviously, for the Shore towns that were fully evacuated, and for people who could not return to their own homes because of extensive damage, it was much harder for routines to reassert themselves. A Monmouth University/*Asbury Park Press* survey of people who had been displaced for at least a month found that eighteen months later, more than one in three (34 percent) reported having a "mental health condition or emotional problem" that had kept them from normal activities for at least one day in the previous month.[5] Nearly one in twelve (8 percent) reported they had spent sixteen to thirty days in the previous month plagued by mental health or emotional problems. The survey used a scale of psychological distress that measured how frequently individuals felt nervous, hopeless, worthless, restless, or fidgety, so depressed that nothing could cheer [them] up, and that everything was an effort. The responses of nearly one out of every four (24 percent) indicated "serious psychological distress," and another 23 percent reported symptoms of mild to moderate distress, in comparison with just 4 percent and 10 percent, respectively, for other New Jersey residents. The survey also found that among those who remained displaced from their pre-Sandy homes, 63 percent reported at least mild distress, in comparison with 35 percent of those who had returned to their pre-Sandy homes.

Notably, Sandy's effects spread well outside the communities that suffered the greatest damage and disruption. Public Service Gas & Electric (PSE&G), the largest electric utility in the state, reported that 77 percent of its 2.2 million customers had lost power at least temporarily during the storm. Jersey Central Power & Light (JCP&L), the state's second-largest electrical utility, which services all of Monmouth County and most of Ocean County, reported more than 930,000 customers without power on October 30. Atlantic City Electric, which serves most of South Jersey outside of the Philadelphia suburbs, including southern Ocean County and all of Atlantic and Cape May Counties, reported that at least 220,000 customers lost power during Sandy.[6] In other words, the electrical grid in the entire state was affected, as were the grids of neighboring states—*Time* magazine reported at the end of November that 8.1 million homes in seventeen states had lost power as a result of the storm.[7] Power outages persisted in some locations. Four days after the storm, 1.2 million PSE&G customers remained without electricity.[8] Ten days later, 33,600 customers who lost power during Sandy still had no power, and another 21,700 had lost power as a consequence of a nor'easter that hit the region on November 7.[9]

The collapse of the electric grid was caused mainly by the loss of connecting wires as a result of high winds, which blew down wires, snapped utility poles, and felled trees that brought wires down with them. The leafy suburbs of New

York and Philadelphia transformed into a minefield of downed trees and live wires. In my neighborhood, forty-five miles inland, an entire row of utility poles snapped like matchsticks and dangled ominously along the major street that leads to my son's day care center. PSE&G reported that within two weeks of the storm, it had replaced twenty-five hundred poles and one thousand transformers, and had cut down forty-one thousand trees.[10] Utility crews worked sixteen-hour days, assisted by more than twelve thousand utility workers from forty-one other states and the Province of Ontario.[11]

The combination of falling trees and problems with electricity was more than an inconvenience for residents. In its coffee-table book about Sandy, the *Star-Ledger* reported thirty-eight deaths in New Jersey linked to the storm and its immediate aftermath, and these were spread among fifteen counties.[12] There were more deaths in both Essex (five) and Middlesex (six) Counties than in any of the four coastal counties, which recorded only seven deaths total (four in Ocean, two in Atlantic, and one in Cape May). Falling trees accounted for the largest number of deaths (nine). Although flooding in Shore communities did result in five deaths, an equal number of people fell in their homes during the power outages or were asphyxiated by generators. Four additional deaths were attributed to fires caused by generators or candles during the blackout, and another two were linked to the failure of oxygen tanks. In all, lack of power was linked to eighteen of the thirty-eight deaths. The combined human cost of trees and power failure in New Jersey was twenty-seven lives, or

6.2. Tree on electrical wires, Ewing, October 30, 2012. Photo by author.

more than two-thirds of all deaths linked to Sandy in the state; the vast majority of these were outside the four coastal counties (thirty-one of thirty-eight, or 81.6 percent).

The loss of power and the seemingly haphazard way in which it was restored left many residents frustrated. One of the most vocal discontents (with media access) was Mayor Dave Fried of Robbinsville, which is located about thirty miles inland from the coast; half of Robbinsville is served by PSE&G and half by JCP&L. The week after the storm, Fried could be heard on local radio praising PSE&G for its rapid restoration efforts. He was a much harsher critic of JCP&L; at one point, he personally went to the JCP&L substation in his municipality and called a radio station to complain about the lack of workers at the site. Within a month, the mayor had written an op-ed piece in the *Star-Ledger* calling for a strict review of JCP&L's performance, writing that the utility was a "a company more interested in complaining that New Jersey has too many trees than in fixing itself."[13]

Gas stations across the state also reeled from power outages. Stations generally rely on electricity to pump gas from underground storage tanks. Unless a station had a generator, there was no way to refuel cars; moreover, without electricity, stations could not process credit cards. As a consequence, many gas stations remained closed during the first week after the storm. Of those that were able to open, many would only accept cash. As residents began to realize a gas shortage was imminent, they began to line up at those stations that were open. By November 1, a *Star-Ledger* headline warned of "long lines, short tempers," but attributed the "panic" to "herd mentality" rather than actual shortages.[14] Long waits and flare-ups among those waiting in line to get gas compelled first the presence of police officers and eventually, the institution of mandatory rationing in twelve counties. Rationing meant that cars with license plates ending in odd numbers could gas up one day and those with even numbers the next. Although inland portions of the four coastal counties faced some of the greatest actual shortages, only Monmouth was subject to rationing—the other eleven counties were all in the northern and northwestern parts of the state: Bergen, Essex, Hudson, Hunterdon, Middlesex, Morris, Passaic, Somerset, Sussex, Union, and Warren.

Price gouging and scams also became part of the post-Sandy social disorganization. The Office of the Attorney General released a press statement on October 31 indicating that it had already received "approximately 100" complaints of gouging for gas, generators, and hotel rooms. Two days later, this had risen to "more than 500" complaints and then over a thousand by the end of November.[15] On the storm's anniversary, the Division of Consumer Affairs reported that it had investigated more than two thousand allegations of gouging, although only twenty-seven lawsuits were actually filed, mostly against hotels and gas stations.[16] State officials began warning of home repair and charity scams as early as November 1, urging residents to "be extremely wary" of "fly-by-night opportunists who may have come in from out of state—or those who may live locally but lack the skills and honesty you need for a significant repair job."[17] Within three months, the New Jersey Division of Consumer Affairs was investigating "dozens" of charities that had solicited donations for Sandy victims, and Acting Director Eric T. Kanefsky was quoted as explaining,"We know charlatans will seek to use tragic events to enrich themselves—and because even honest organizations may need guidance in order to come into full compliance with our consumer protection laws."[18]

In short, in addition to the physical devastation of flooded neighborhoods, Garden State residents faced blackouts, home repairs, dangerous trees, gas shortages, fraying tempers, price gouging, scams, and a variety of other departures from normal life. Tens of thousands were displaced from their homes and were living in shelters and hotels or were doubled up with friends or family. Virtually all 8.8 million residents of the Garden State had their daily routines disrupted, as workplaces, schools, government, and transportation systems slowly resumed normal operations after statewide shutdowns.

These changes were compounded by the sadness many felt as images of their beloved Shore landmarks began to circulate. As I discussed in the second chapter in this book, most New Jersey residents have strong emotional ties to the Shore; these ties are often linked with family, friends, youth, and free time at the Shore. The destruction of the Shore represented an emotional loss for people who live far away from the Shore communities themselves. Upon viewing pictures of the Star Jet roller coaster lying in the surf off of Seaside Heights, my husband lamented that our children would not be able to ride the same rides that we did as children; the continuity of Jersey Shore culture was lost when these attractions were destroyed. This sentiment was widely echoed in social media, where overwhelming amounts photographic evidence of the destruction and sad narratives about lost local landmarks were shared on sites such as Jersey Shore Hurricane News (JSHN), which quickly came to serve as a sort of memorial site for the damaged and destroyed cultural landscape even while helping direct the flow of information and volunteers in the region.[19]

Ontological Security in a Risk Society

For some Shore residents, Sandy punctuated the end of an era: "The Storm That Forever Changed New Jersey."[20] But for most of us, it was less the apocalypse than a warning shot across the bow of modernity: we are vulnerable. Rutgers sociologist Lee Clarke, who has built his career around the examination of risk and disasters, explained in the winter 2013 issue of *Contexts*, a publication of the American Sociological Association, that "one thing Sandy was not was a worst case" because it was well anticipated and caused relatively few human casualties.[21] While Sandy was the largest storm to affect the Jersey Shore in a half century, it was hardly unprecedented. The technologies and emergency plans that had been put into place, and tested the year before with Hurricane Irene, largely worked. Initial recovery and even longer-term recovery efforts went fairly smoothly, particularly in comparison with the debacle of Hurricanes Katrina and Rita in 2006. Emergency management officials up and down the coast rightfully congratulated themselves on having averted the worst. Even so, a Rutgers-Eagleton survey conducted a year after the storm found that only 26 percent of the state's residents believe that the state was "back to normal," and this figure dropped to only 14 percent among Shore residents.[22]

When a disaster affects entire communities, both the individuals and the community itself are unsettled. After interviewing the victims of the Buffalo Creek floods in West Virginia, Kai Erikson called the disruption of social systems "collective trauma" and noted that it compounded individual trauma and loss.[23] Individuals were psychologically scarred from the loss of lives, security, and their homes, but they also suffered from having their social networks splintered, from having routines disrupted, and from the presence of physical reminders of the disaster itself. Members of the community showed high levels of depression, mistrust, and self-destructive behaviors. Sociologists Duane A. Gill and J. Steven Picou found similar results following the *Exxon Valdez* disaster in Alaska, which decimated rural fishing communities.[24] Subsequent research on Hurricane Katrina has confirmed that in addition to individual psychological stress, individuals in disrupted communities also suffer from a malaise linked to a loss of both their community and their assumptions about the way the world works.

Sociologist Anthony Giddens describes how the lack of feeling "back to normal" wears on individuals. He describes contemporary social life as one of deep embeddedness in "abstract institutions" that make modern living possible but of which individuals have limited understanding, choice, or control. Abstract systems rely on experts for their operation, with lay individuals rarely understanding their intricate inner workings. Trust in these systems is thus not generally the result of the rational individual calculations, but of experience with experts who represent those systems, and the familiarity, reliability, and continuity of day-to-day living. Giddens also points out that there are few alternatives to trusting abstract systems, so "trust is much less of a 'leap to commitment' than a tacit acceptance of circumstances in which other alternatives are largely foreclosed."[25] When these abstract systems fail to produce a familiar, reliable outcome, they not only produce mistrust of the systems themselves but also generate feelings of angst and dread as people face evidence of their own vulnerability within those systems.

According to Giddens, our modern "risk profile" is not just about how we respond to natural disasters but about how we respond to "socialized nature" or "the altered character of the relation between human beings and the physical environment."[26] Modern humans don't just interact with nature, but with a series of abstract institutions that moderate and regulate this interaction. Thus, experience with nature, and even experience with "natural" disasters, reflects our interactions with and feelings about abstract social systems.

At the most fundamental level, this involves the physical nature we experience: the Jersey Shore is essentially an engineered coastal environment, with roots in the physical landscape and the political and technological knowledge that has replenished, stabilized, bulkheaded, privatized, developed, and managed that landscape. At only a few places can a modern individual experience the Jersey Shore in its "natural" state. Even these places, such as Sandy Hook and Island Beach State Park, are highly managed, somewhat idealized restorations of what we collectively believe the natural shoreline should look like. These landscapes are maintained by the artificial stabilization of the Sandy Hook spit and the Barnegat Inlet respectively, among other environmental management decisions.

Modern experience with environmental risks is also heavily influenced by our trust in abstract systems. For example, we determine how dangerous a storm is by listening to meteorologists and emergency management officials. However, interactions with these experts can actually lead to *doubting* their expertise. Lee Clarke found that this is because people think about risk to themselves and the people they love in terms of *possibilities*, while experts think about and minimize risk through *probabilities*.[27] What does it matter if there was only a one in one hundred chance that floodwaters would destroy my home if mine is the one house that gets destroyed? Experience with events also can erode confidence and trust in the systems. For example, the slow and inefficient evacuation of Absecon Island in advance of Hurricane Irene in 2011 caused the mayor of Atlantic City to assume that many residents would not evacuate for Sandy, even through multiple experts had issued dire warnings against remaining on the barrier islands.

Despite our tenuous trust in expert-run abstract systems, we understand that there are myriad environmental and ecological risks that could potentially affect us every day; even so, we have become numb to them because our daily experiences reinforce ontological security, and to do otherwise would lead to madness. When disruptions to that status quo like Sandy occur, *some* people may be overcome with angst and dread, but *most* adapt because modern society simply cannot function without abstract systems in place.[28] Giddens identifies four possible forms of adaptation: "pragmatic acceptance," which leads people to focus on the elements that one can control in their daily lives, including helping friends and neighbors; "sustained optimism" based on faith that all problems can be overcome, especially with advanced science and technology, such as ecological modernization theory predicts; "cynical pessimism," which acknowledges the fundamental flaws of modern social organization but tempers the outlook with a fatalistically bleak humor; and "radical engagement" against the source of the threat. Sandy teaches us that the first three adaptations, all of which allow the return to normal routines and get back to mostly ignoring abstract systems, are the most common.

Ontological security becomes paramount as the modern world is increasingly characterized as what German sociologist Ulrich Beck calls a "risk society."[29] To Beck, as material scarcity becomes less of a pressing concern (as is the case in most advanced industrial economies like the United States), exposure to risks becomes more important for understanding social organization.[30] In other words, when most people no longer have to mobilize collectively to meet basic needs of food, shelter, and the like, they can turn their political attention to minimizing risks. Thus, the main social debates in a risk society become about potential futures rather than existing material conditions: What if government gets too big? What if climates become destabilized? What if immunizations have disastrous side effects?

Risk controversies help undermine scientific rationality, as science, which is predicated on empirical evidence, simply cannot provide unequivocal security from modern risks. Although risks are often mathematically predicted, no risk calculation can possibly consider all of the potential variables that will produce a specific outcome. At best, probabilities can predict the most likely outcome. Scientists themselves have no problem with this. The failure to predict a specific outcome is both anticipated in the logic of probabilities and is actually desirable to advance scientific knowledge. Unfortunately, outside the community of scientists (and sometimes within it), probabilities are often misunderstood, as when people assume that the longer they play at a slot machine, the greater their chances of winning a jackpot.[31] In a more relevant example, FEMA's elevation maps may reflect the best prediction given what scientists know about the geophysical risks to the Shore region, but this doesn't mean that structures outside that section of the map are safe from future storms, nor does it mean that any structure within the map is guaranteed destruction.

Because science is unable to provide true security from risks, people begin to question if science and scientists deserve the authority they claim. Empirical experiences may be used as evidence that refutes the claims made by scientists—if two homes were predicted to be at equal risk and one was destroyed and the other was not, how are probabilities helpful? What about when one house that had low predicted risk was destroyed, and one with high risk was untouched? Anecdotes may be interpreted to mean that the probabilities were "wrong," which calls into question future risk calculations made by scientists. This is further complicated by a somewhat unusual culture in the United States that is more likely to doubt science, in no small part because of a concentrated effort to obfuscate or deny scientific predictions. This is clearly demonstrated by the organized effort to deny and discredit climate change science, which has found more traction in the United States than anywhere else in the world.[32]

The legitimacy of scientific knowledge is also called into question when the science is inconclusive, or when science suggests competing solutions. As previously discussed, Bay Head has decided to extend its sea wall as a means for reducing future risks, while Mantoloking rebuilds its dune line. The sea wall *seemed* to have protected Bay Head during Sandy, but dunes worked elsewhere and are recommended by experts such as the DEP and the Army Corps of Engineers, who are supposed to have scientific knowledge about such matters. Likewise, should elevation of structures be linked to existing hazards, or should they take into account projected sea level rise, and if sea level rise is considered, which model of sea level rise should be used? Even if one accepts that science has the best solutions, there is no clear way to determine which model is best. In this way, scientific predictions become claims that must compete in a market of other claims, many of which may be more appealing to individuals for nonscientific reasons. For example, elevating existing structures above the minimum requirement for flood insurance requires individuals to pay more; higher elevations may reduce risk, but will the reduced risk merit the increased cost?

Beck also alerts us to the fact that even when scientific assertions are not in doubt, modern humans face multiple simultaneous risks that must be socially prioritized. The DEP's own guiding documents require it to balance disaster mitigation with ecological needs and social and especially economic concerns. In any given social circumstance, there are a limited number of resources and there are often competing and compelling social worries that carry greater weight than more amorphous future risks—even if those risks may be catastrophic. Which is of more pressing concern: getting displaced people back into their homes or elevating those homes? Reopening the Mantoloking Bridge or considering the long-term viability of permanent infrastructure on barrier islands? Rebuilding boardwalks or rebuilding homes? Here, the perception of risk and the ability to assert that perception are more important than the risk itself. In addition to getting the economy moving again, people want desperately to get back to the routines that make them feel safe about living in a structurally risky society. The restoration of ontological security, that is, the appearance that the risky systems on which we depend are not risky at all, is central to this return to normal. Reducing the perception of risk is actually more important after a disruption to ontological security than actually reducing risk.

Scientific reasoning, which has been undermined by the processes discussed above, no longer provides a clear path forward, so responses to events like Sandy arise from social, rhetorical, and nonrational reasoning.[33] All of these typically privilege the interests of people with advanced levels of education, wealth, and access to the media. The ability to assert their interests in the face of disasters like Sandy helps explain why the federal government authorized $60 billion in the Sandy Supplement in March 2013,[34] long before the results of even the revised FEMA maps were available. It is virtually impossible for these funds to be distributed solely based on scientific estimates of risk, but this did not stop the flow of public funds.

This is not to suggest that risks are not real. Physicist Neil de Grasse Tyson has been quoted as saying, "The good thing about science is that it's true whether or not you believe in it." The same is true for the risks associated with modern living; risks are inherent and central to modern social organization. People felt the need to return to normal after the storm, and socially, this return to normal could be framed as economic and moral imperatives. People want and feel entitled to access to environmental amenities on the Jersey Shore, both as residents and visitors. Technological and governmental solutions were made available, and people have moved or are in the process of moving back into harm's way. As engineers, officials, residents, and volunteers rebuild the Shore, they also lull us back into complacent ontological security, so we go back to ignoring the precariousness of how we have chosen to live.

Conclusion: Everything That Dies, Some Day Comes Back

The Jersey Shore will never be the same as it was before Sandy, but it also won't be a vastly different place. Human history abounds with massive disruptions; this has not yet meant the end of the world as we know it. Sandy completely destroyed the built environment in some areas, wiping out individual lives, memories, and property. Yet the storm's impact on broader social organization was limited. People will continue to interact with the Jersey Shore principally as a recreational landscape, with some new rides, new shops, and new boardwalks. Poorer residents and their homes will continue to be replaced by more affluent residents and larger, elevated homes. Flood insurance premiums will go up. Infrastructure will be "hardened" against future storms. All of these things were in process before Sandy; the storm may have sped up the process but it did not derail it.

Despite the proliferation of "Stronger than the Storm" and "Restore the Shore" paraphernalia within the state, the Jersey Shore is back off the national cultural radar. A flurry of national news stories around the one-year anniversary of the storm highlighted the recovery effort. On October 29, 2013, Governor

Christie appeared on a five-minute segment on CBS's *This Morning* show, but the discussion shifted away from recovery after just two minutes, focusing instead on partisan politics and Christie's weight loss. In August 2014, the Weather Channel's *Hurricane 360* series reenacted of some of the more harrowing personal narratives during Sandy in an episode titled "Horror on the Jersey Shore," but the reconstruction effort was not its focus.

How people think about the Jersey Shore may or may not have changed. In their letter to potential tourists in the state's official 2014 Travel Guide, Governor Christie and Lieutenant Governor Kim Guadagno wrote: "This past year, we witnessed the strength, compassion, and generosity of our residents in the aftermath of Superstorm Sandy. New Jerseyans came together to help rebuild our communities, reopen our businesses and restore our beautiful Jersey Shore. It's with great pride that we can let the nation know that New Jersey is indeed open for business!"[35] The 144-page guide does not mention the storm again, even in its profile on boardwalks. Instead, it references the "action-packed, mile-long Seaside Height boardwalk, one of the most popular and most visited in the state. A magnet for young people, it features the Casino and Funtown piers."[36] Never mind the ongoing reconstruction at the Casino Pier and the fact that the Funtown Pier simply no longer exists.

Shore residents, however, are well aware that their world has changed dramatically. Towns and neighborhoods on the Barnegat Peninsula are the most affected, with many residents and second-home owners persisting in the rebuilding process. What is being rebuilt is often not similar to what was destroyed; buildings are larger, more modern, and elevated. Open lots remain where houses were swept away or demolished. Many damaged homes remain in place while residents seek additional funding to rebuild. Some homeowners walked away, and municipal governments are forced to condemn the properties.

The physical appearance of the Shore has also changed dramatically. Homes are gone. Businesses are gone. In Belmar, all structures east of the boardwalk were demolished. In Seaside Heights, the Casino Pier remained an active construction site. At the end of the summer 2015 season, Seaside Park had not recovered from the fire—the pier is gone, all of the boardwalk businesses are gone. Only the Sawmill restaurant remains, touting "a (temporarily) unobstructed view of the beautiful Seaside Park beach and Atlantic Ocean, while the iconic boardwalk awaits rebuilding."[37] The view east from the Mantoloking Bridge still looked directly onto the ocean, now barely visible behind a massive dune. Three years out, many of the lots in the oceanfront blocks of Ortley Beach remain undeveloped. Other blocks have a jack-o'-lantern appearance, with a combination of low homes (some of which remain boarded up), empty lots, and elevated homes. Communities flooded by bays and rivers have come back a little faster, but only because there were more year-round residents and more immediate access to begin rebuilding.

The Shore's tourism economy has partially rebounded. The public emphasis on boardwalks allowed most of these attractions to accommodate visitors in 2013 and even more in 2014. The focus on rebuilding the boardwalks was great "public relations," but the benefits of this restoration went mainly to the region's growth machine interests. Ocean County's barrier beach towns were most affected. Both Monmouth and Cape May County beaches probably benefited from the slow rebuilding in Ocean County, as people swarmed to places like the Wildwoods and Asbury Park.[38] My family spent one frustrating Sunday in summer 2014 driving around Asbury Park, unable to find a place to park—even lots charging up to twenty dollars had lines of cars waiting to get in. Throughout the summer, food vendors on the Asbury Park boardwalk had lines dozens of people deep. The beach was filled blanket-to-blanket. Waits at restaurants in Asbury, Belmar, and other Monmouth County Shore towns were commonly over an hour.

Ironically, the news out of the Shore at the end of the summer 2014 often featured the troubled economy of Atlantic City. By September, four casinos had shuttered or were slated to close, among them the luxury Revel Resort, which had opened just two years before. The proximate cause for the failure of the casinos was the legalization of gambling in neighboring states (most critically, Pennsylvania). But the strength of the Shore economy elsewhere suggests that the problem with Atlantic City is not just the casinos, but the failure to provide the type of environment that makes money on the Jersey Shore: safe, family-oriented beach time.

Families are returning to the Shore, but many continue to struggle with recovery efforts. Primary-home owners have received funds from FEMA, HUD, and the state, as well as private insurance, but in few cases were the reconstruction funds sufficient to cover the economic costs of displacement, reconstruction, and mitigation. Moreover, the lack of funds available to second-home owners has reduced the number of people in residential neighborhoods. Families who relied on Shore rental properties for income face a triple burden: restoration without assistance because most public programs restrict aid to second homes; the increased cost of flood insurance since the HFIAA does not delay or reduce flood insurance costs to second-home owners; and the inability to earn rental income.

It is unclear how much Sandy will change the social environment along the Jersey Shore. Although residents, volunteers, and officials all say the storm has brought communities together, they also mention increases in poverty and economic struggles. Mental health concerns abound. The cost of rebuilding and flood insurance has forced longtime residents and homeowners to relocate permanently, eroding social networks that span decades. Many expect the demographic profile of residents to change; there will be fewer elderly residents, fewer renters, fewer middle-income families.

The government has become a constant presence. Federal agencies like FEMA and HUD, as well as the Army Corps of Engineers, have become part of the everyday language and reality for residents. These are complemented by various state agencies, including the DEP and the Department of Transportation, which have various highly visible recovery projects throughout the Shore region. Elevation and flood insurance remain hotly contested issues; discussion of either can evoke tears and anger from near-strangers.

The government has taken an activist stand on preparation for the next storm. The Borough of Harvey Cedars on Long Beach Island, when seeking to create protective dunes along its shoreline, condemned oceanfront land that had been owned by the Karan family since 1973. The borough offered the Karans $375,000 in compensation, but the Karans sued for $500,000, citing a real estate agent who had assessed their property as being devalued by this amount as a result of the loss of ocean views.[39] Prior to Sandy, the initial trial court and subsequent appeal upheld the Karans' claims, principally by disallowing jurors from balancing the devaluation of the lost view with the value of the storm protection afforded by the dunes. However, the New Jersey Supreme Court, which heard the case in 2013, found that these instructions to the jury were inappropriate and that the value of storm protection must be considered when determining "just compensation" for oceanfront property seized through eminent domain. This decision, along with mounting public frustration and harassment of oceanfront homeowners who had not signed easements for dune construction, led the Karans to settle with Harvey Cedars for $1 in September 2013.[40] This decision cleared the way for the development of a continuous dune berm elsewhere along the Jersey Shore.

There is little to suggest, however, that people have begun to question the way that humans interact with the nonhuman environment on the Jersey Shore. Tourism remains the main focus of the region's political and economic interests. People still want permanent occupation of the Shore, even in the most precarious locations. And they hope the Shore will remain locally oriented and economically accessible—full of history and memories for "average" people—not glitzy resorts for the elites, like the failing Atlantic City casinos. The government has largely acted in response to those desires, even at time sacrificing long-term public safety and environmental protection.

Of course, it is just a matter of time before another storm devastates the region. David A. Robinson, New Jersey's state climatologist, has stated in multiple venues that historical weather data demonstrate that the state's climate is getting warmer and wetter; perhaps more importantly, sea level is rising.[41]

Although storms like Sandy still cannot be predicted with any certainty, storm surges will cause greater damage as sea levels rise. As residents' perception of safety increases through technological fixes like dune berms and elevation, it is plausible that fewer will evacuate, particularly as the memory of Sandy recedes over time.

Three years after the storm, "recovery" is evident, albeit uneven. Most of the boardwalks along the Shore have been replaced, often with improved facilities. New homes have been constructed, old homes remodeled and elevated. Ongoing work has created traffic woes for the Barnegat Peninsula, but improvements to State Highway 35 and the bridges at Mantoloking and Toms River are designed to harden this infrastructure against future storms. Many lots remain empty where the waters washed away homes; many homes still need repairs to make them habitable. Residents have forgone other things to pay for repairs and have found themselves with new debt loads. They have been forced to move to other communities while waiting for insurance money or repairs. Tenants have found the rental market severely curtailed and increasingly expensive, both because of the loss of units and because so many residents are renting while their primary homes are being repaired. Some residents have given up and moved away, selling heirloom properties to people better able to absorb the costs of remediation and mitigation. The modest Jersey Shore bungalow is giving way to larger, modern houses on stilts. New faces and new buildings have undoubtedly changed some the character of the Jersey Shore.

This book has outlined how the damage caused by Sandy along the Jersey Shore followed foreseeable patterns, and how reconstruction efforts following the storm were equally predictable. Environmental sociology can provide explanations for why the storm caused so much property damage, as well as why reconstruction puts that property back in harm's way. While perhaps not logical, it is not a mystery why people don't want to change. Humans are not just biological creatures, and our environment is not just our habitat; it is the manifestation of our social world. Our relationship with the environment reflects vested economic interests and deep emotional ties; the Jersey Shore was built on economics, politics, and social relationships. And these social structures and social networks are all stronger than any single storm, even one as destructive as Sandy.

At the same time, resistance to change is hard. As much as individuals adapt to new circumstances, societies must adapt to new conditions. In New Jersey as elsewhere, we acknowledge that storms like Sandy are inevitable and are likely to become more common. Decisions made in the Sandy recovery period have only partially this into taken into account. Strong social and political pressure to keep the Shore accessible to nonelites, while commendable on one hand, shifts the actual cost of coastal development to people who live far from the ocean, and generally fails to reduce risk to those who were most vulnerable in Sandy. One of the most important lessons that we have learned from Sandy is that there is limited political will to entertain alternatives to rebuilding the Shore as a place of dense, human occupation with easy access to recreational amenities. In many ways, this dooms us to a predictable pattern of future destruction followed by reconstruction followed by later destruction, ad infinitum. And things can get much worse: if a storm of comparable power strikes the less economically resilient and less politically influential Atlantic City, the result might look more like Katrina than Sandy.

I very much want my kids to once again ride the carousel at the Funtown Pier.[42] It is part of my personal history and part of a cultural tradition that I share with millions of other people across time and space. But I don't want this future to be built on exclusion of others or a mountain of public debt accumulated from an endless cycle of expensive coastal storms. Although most of the decisions about recovery on the Jersey Shore have already been made, Sandy continues to provide an opportunity for us—in New Jersey and beyond—to think more carefully and deliberately about the multifaceted ways in which human society interacts with its environment. By examining the Jersey Shore, both before and after Sandy, I suggest that we need to consider more than free markets and engineering when we make social decisions. To this end, this book has attempted to find ways that environmental sociology can contribute to a process through which we can become not just stronger than a storm, but better suited to live on this planet.

NOTES

PROLOGUE

1. All accounts and quotations from this section are taken from interviews conducted with Ocean County residents, second-home owners, volunteers, and officials as part of the Hurricane Sandy Oral History Project in fall 2013, led by Dr. Matthew Bender of the College of New Jersey and available at: https://history.tcnj.edu/program-information/hurricane-sandy-oral-history-project. Although oral histories were collected from individuals representing nine municipalities, this project focused primarily on two of the hardest-hit areas of the Jersey Shore along the Barnegat Peninsula: Mantoloking/Bay Head/Point Pleasant Beach and the Toms River beaches/Seaside Heights. Damage from Sandy was much more widespread on the Barnegat Peninsula, but these narratives provide a compelling window into the experiences of Shore residents before and after Sandy.
2. The Toms River Police Department estimated that 95 percent of residents in the evacuation zones did vacate; the Mantoloking Police Department reported only fifteen people in the entire town chose not to evacuate.
3. The Barnegat Peninsula, which extends from Point Pleasant and Bay Head south to the tip of Island Beach State Park at the Barnegat Inlet, is often referred to locally as a barrier island.

CHAPTER 1 THE INEVITABLE SANDY

1. Eric S. Blake, Todd B. Kimberlain, Robert J. Berg, John P. Cangialosi, and John L. Beven II, "Tropical Cyclone Report: Hurricane Sandy (AL182012) 22–29 October 2013," National Hurricane Center, Miami, Florida, 2013, http://www.nhc.noaa.gov/data/tcr/AL182012_Sandy.pdf, 120.
2. Ibid., 3.
3. Ibid., 114.
4. John Erdman, "Superstorm Sandy: A Giant Circulation," Weather Channel, November 29, 2012, http://www.weather.com/news/weather-hurricanes/sandy-another-giant-storm-20121027. Also see Rob Gutro, "Hurricane Sandy (Atlantic Ocean)," National Aeronautics and Space Administration (NASA), March 7, 2013. http://www.nasa.gov/mission_pages/hurricanes/archives/2012/h2012_Sandy.html.
5. Blake et al., "Tropical Cyclone Report," 10.
6. Ibid., 17.
7. Ibid.
8. The "first wave" of US environmentalism took place at the end of the nineteenth century and beginning of the twentieth century and is associated with important legacies such as national parks and national forests. See Riley Dunlap and Angela Mertig.

American Environmentalism: The U.S. Environmental Movement, 1970–1990 (New York: Routledge, 1992).

9. With apologies to the New Jersey Devils, hockey loyalties are divided in the state between the Devils, the New York Rangers, and the Philadelphia Eagles.
10. The complete citations for theoretical works and concepts mentioned in this section can be found in the chapters that follow, as well as in the bibliography.

CHAPTER 2 THE SHORE OF MEMORIES

1. Coastal Research Center, Richard Stockton College of New Jersey, "NJ Shoreline Protection and Vulnerability," 2014, http://intraweb.stockton.edu/eyos/page.cfm?siteID=149&pageID=4.
2. Thomas Belton, *Protecting New Jersey's Environment: From Cancer Alley to the New Garden State* (New Brunswick, NJ: Rutgers University Press, 2011). See especially chapter 2.
3. There is also a small portion of Cape May that can be physically characterized as a headland, but its contemporary social use is similar to other parts of Cape May.
4. New Jersey Department of Community Affairs, "New Jersey Community Development Block Grant Disaster Recovery Action Plan State Fiscal Year 2013," Department of Community Affairs, Trenton, 2013, http://www.nj.gov/dca/divisions/dhcr/rfp/pdf/cdbg_dr_plan_rfp_gl.pdf.
5. United States Environmental Protection Agency, National Estuary Program, "Effects of Artificial Shorelines," http://water.epa.gov/type/oceb/nep/upload/2009_05_28_estuaries_inaction_Adaptable_BarnegatBay.pdf.
6. Franklin Ellis, *History of Monmouth County, New Jersey* (Philadelphia: R. T. Peck & Co., 1885), 7, https://ia802704.us.archive.org/16/items/historyofmonmoutooellis/historyofmonmoutooellis.pdf. Lorraine E. Williams and Ronald A. Thomas, "The Early/Middle Woodland Period in New Jersey: ca. 1000 BC–1000 AD," in *New Jersey's Archaeological Resources from the Paleo-Indian Period to the Present: A Review of Research Problems and Survey Priorities* (Trenton: New Jersey Department of Environment Protection Office of Cultural and Environmental Services, 1982), 125.
7. Anne Schillingburg and F. Alan Palmer, "The Unalachtigo of New Jersey: The Original People of Cumberland County," http://www.co.cumberland.nj.us/filestorage/163/233/239/unalachtigo.pdf.
8. C. A. Weslager, *The Delaware Indians: A History* (New Brunswick, NJ: Rutgers University Press, 1990).
9. See Williams and Thomas, "Early/Middle Woodland Period."
10. Edwin Salter, *A History of Monmouth and Ocean Counties: Embracing a Geneological Record of Earliest settlers in Monmouth and Ocean Counties and their Descendants. The Indians: Their Language, Manners, and Customs. Important Historical Events: The Revolutionary War, Battle of Monmouth, The War of the Rebellion: Names of Officers and Men of Monmouth and Ocean Counties Engaged in It, Etc., Etc.* (Bayonne, NJ: E. Gardner & Son, Publishers, 1890), 5.
11. Ibid., 7.
12. Ibid.
13. Ellis, *History of Monmouth County*, 26.
14. Charles A. Stansfield, *A Geography of New Jersey: The City in the Garden*, 2nd ed. (New Brunswick, NJ: Rutgers University Press, 2004), 237–238.
15. Corruption in the Prohibition Era is the subject of Nelson Johnson, *Boardwalk Empire: The Birth, High Times, and Corruption of Atlantic City* (Medford, NJ: Medford Press, 2002), which was fictionalized in HBO's series *Boardwalk Empire*. One suggestion for why New

Jersey was more corruptible than other parts of the United States was that immigrants from countries that never banned the sale of alcohol made up a large proportion of residents, and they were less likely to support Prohibition than Protestant, native-born whites. It is also clear that Atlantic City and other New Jersey cities were able to convert bootlegging into impressive profits; this was probably the primary incentive for corruption, especially in the absence of moral condemnation.

16. Bryant Simon, *Boardwalk of Dreams: Atlantic City and the Fate of Urban America* (New York: Oxford University Press, 2004).
17. Jarrett Renshaw, "Gov. Christie Vetoes 'Jersey Shore' Tax Credit," *Star-Ledger*, September 26, 2011.
18. *Sopranos* creator David Chase's HBO's series *Boardwalk Empire* adds to the narrative of the Shore by highlighting the widespread corruption of government in Atlantic City during Prohibition, but this historical drama makes little effort to link itself to the contemporary Shore.
19. Chris Christie, "Governor Christie: There's Nothing More Jersey than the Shore," Governor Chris Christie Press Releases, August 29, 2013, http://nj.gov/governor/news/news/552013/approved/20130829f.html.
20. John Hannigan, *Environmental Sociology*, 2nd ed. (New York: Routledge, 2006).
21. Ibid., 64.
22. Ibid., 70.
23. Ibid., 73.
24. Kari Marie Norgaard, *Living in Denial: Climate Change, Emotions, and Everyday Life* (Cambridge, MA: MIT Press, 2011), 5.
25. Rutgers-Eagleton Poll, "Sandy's Legacy: Climate Change Is Real for New Jerseyans, Rutgers-Eagleton Poll Finds," Eagleton Institute of Politics, April 29, 2013, http://eagletonpoll.rutgers.edu/new-wp/04-29-2013-superstorm-sandy-and-global-climate-change-beliefs/.
26. Julie O'Connor, "One Year after Sandy, Christie Officials Sleepwalking on Climate Change: Opinion," nj.com, October 27, 2013, http://blog.nj.com/perspective/2013/10/one_year_after_sandy_christie.html.

CHAPTER 3 SHORE RESORTS

1. Tourism Economics, *The Economic Impact of Tourism in New Jersey: Tourism Satellite Account Calendar Year* 2012 (Trenton: New Jersey Division of Travel and Tourism, 2012), http://www.visitnj.org/sites/default/master/files/2012-nj-tourism-ei-state-counties-v0701.ppt, 68, 71, 73.
2. Ibid., 42.
3. Ibid., 3.
4. Ibid., 30–31.
5. Ibid., 31.
6. Ibid.
7. Ibid.
8. Ibid., 28
9. New Jersey Casino Control Commission, 2011 *Annual Report* (Atlantic City: New Jersey Casino Control Commission, 2011), http://www.nj.gov/casinos/reports/docs/2011_ccc_annual_report.pdf, 45.
10. Tourism Economics, *Economic Impact of Tourism*, 8.
11. Ibid., 10.

12. Ibid., 13
13. Ibid., 24. This source reported 81.3 million visitors to New Jersey in 2012; 30 million is estimated as a percentage of this number based on the lowest percentage share of the tourism industry noted in table 3.1 associated with Shore Counties (41.84 percent).
14. Scott Stump, "Jersey Shore Reopens 7 Months after Sandy: We're 80% There, Says Christie," *Today News*, NBC, May 24, 2013, http://www.today.com/news/jersey-shore-reopens-7-months-after-sandy-were-80-there-6C10052450.
15. David Carr, "Summer on the Jersey Shore, After the Storm," *New York Times*, August 2, 2013, http://www.nytimes.com/2013/08/04/travel/summer-on-the-jersey-shore-after-the-storm.html?_r=0.
16. Patrick McGeehan, "Boardwalk Fire Ruled Accidental, with Hurricane as a Factor," *New York Times*, September 17, 2013, http://www.nytimes.com/2013/09/18/nyregion/jersey-shore-boardwalk-fire-is-ruled-to-be-accidental.html.
17. Harvey Molotch, "Oil in Santa Barbara and Power in America," *Sociological Inquiry* 40 (1970): 131–144.
18. Thomas K. Rudel with Bruce Horowitz. *Tropical Deforestation: Small Farmers and Land Clearing in the Ecuadorian Amazon* (New York: Columbia University Press, 1993).
19. Diane C. Bates, "The Barbecho Crisis, la Plaga del Banco, and International Migration: Structural Adjustment in Ecuador's Southern Amazon," *Latin American Perspectives* 24, no. 3 (2007).
20. See William R. Freudenberg and Robert Gramling, *Blowout in the Gulf: The BP Oil Spill Disaster and the Future of Energy in America* (Cambridge, MA: MIT Press, 2010), for a chilling narrative of how this works in the oil and gas industry.
21. Bryant Simon, *Boardwalk of Dreams: Atlantic City and the Fate of Urban America* (New York: Oxford University Press, 2004).
22. John Bellamy Foster, *Ecology Against Capitalism* (New York: Monthly Review Press, 2002).
23. Ron Cook, Diane Campbell, and Caroline Kelly, "Survival Rates of New Firms: An Exploratory Study," *Small Business Institute Journal* 8, no. 2 (2012), http://www.sbij.org/index.php/SBIJ/article/viewFile/147/91, 39.
24. Christopher Sheldon, "Long Branch Beach Badge Sales Total Over $1 Million," *Long Brach (NJ) Patch*, July 25, 2012, http://longbranch.patch.com/groups/summer/p/long-branch-beach-badge-sales-total-over-1-million.
25. Chris Christie, "Governor Christie Announces Aggressive Plan to Strengthen and Rebuild Jersey Shore's Sandy-Damaged Route 35," Governor Chris Christie Press Releases, February 19, 2013, ttp://nj.gov/governor/news/news/552013/approved/20130219a.html.
26. New Jersey Department of Transportation, "FY 2012–2021 Electronic Statewide Transportation Improvement Program (e-STIP)," DoT, 2012, http://njdotestip.njit.edu/ESTIP/WebTelus/Login:LoginPublic.
27. "Map of the Rail Roads of New Jersey and Parts of Adjoining States 1870," Rutgers Cartography Lab, http://mapmaker.rutgers.edu/HISTORICALMAPS/RAILROADS/NJ_RR_1870.jpg.
28. Franklin Ellis, *History of Monmouth County, New Jersey* (Philadelphia: R. T. Peck & Co., 1885), 756–757, https://ia802704.us.archive.org/16/items/historyofmonmoutooellis/historyofmonmoutooellis.pdf.
29. Ibid., 758.
30. Ibid., 759.
31. Ibid.

32. City of Cape May, "Cape May History," 2009, http://www.capemaycity.com/Cit-e-Access/webpage.cfm?TID=103&TPID=10704.
33. John T. Van Cleef and J. Brogsnard Betts, "Map of the Rail Roads of New Jersey, 1887," Rutgers Cartography Lab, http://mapmaker.rutgers.edu/HISTORICALMAPS/RAILROADS/RR_of_NJ.jpg.
34. The bankruptcy of multiple casinos, including the new luxury resort, Revel, in 2014 cast long-term doubts on the viability of Atlantic City's gambling economy.
35. Matthew R. Linderoth, *Prohibition on the North Jersey Shore: Gangsters on Vacation* (Charleston, SC: History Press, 2010), 16.
36. Ibid., 60.
37. Emil R. Salvini, *Boardwalk Memories: Tales from the Jersey Shore* (Guilford, CT: Globe Pequot Press, 2006), 6.
38. Ocean Grove was compelled to stop gating Main Avenue by the New Jersey Supreme Court in 1979, but it continues to allow no alcohol sales.
39. Palace Museum Online, "1888," Palace Museum Online, 2005. http://www.palaceamusements.com/1888.html.
40. Palace Museum Online, "1902: Eyewitness to History," Palace Museum Online, 2005, http://www.palaceamusements.com/1902.html.
41. Palace Museum Online, "Middle Years, 1955–1956," Palace Museum Online, 2005, http://www.palaceamusements.com/1955_1956.html.
42. Daniel J. Wolff, *4th of July, Asbury Park: A History of the Promised Land* (New York: Bloomsbury Publishing, 2005).
43. Bonnie J. McCay, Debbie Mans, Satsuki Takahashi, and Sheri Seminski, *Public Access and Waterfront Development in New Jersey: From the Arthur Kill to the Shrewsbury River* (Keyport: NY/NJ Baykeeper, 2005), http://nynjbaykeeper.org/wp-content/uploads/2013/05/Long_Branch_Final.pdf.
44. "Pier Village Long Branch," accessed January 28, 2015, http://www.piervillage.com/
45. Jason George, "Testing the Boundary Lines of Eminent Domain; Long Branch Wants to Seize Old Homes to Make Room for New Ones," *New York Times*, March 31, 2004, http://www.nytimes.com/2004/03/31/nyregion/testing-boundary-lines-eminent-domain-long-branch-wants-seize-old-homes-make.html.
46. MaryAnn Spoto, "Long Branch Homeowners Win Round in Eminent Domain Fight," *Star-Ledger*, August 8, 2008, http://www.nj.com/news/ledger/topstories/index.ssf/2008/08/court_ruling_gives_life_to_lon.html.
47. Michael Rispoli, "Long Branch, Homeowners Settle 5-Year Eminent Domain Dispute," *Star-Ledger*, September 15, 2009, http://www.nj.com/news/index.ssf/2009/09/long_branch_agrees_to_end_push.html.
48. See Richard Florida, *The Rise of the Creative Class and How It's Transforming Work, Leisure, and Everyday Life* (New York: Basic Books, 2002), and Richard Florida, *Cities and the Creative Class* (New York: Routledge, 2005).
49. Sharon Zukin, *Naked City: The Death and Life of Authentic Urban Places* (New York: Oxford University Press, 2011).
50. Census population of Asbury Park: 1960: 17,366; 1970: 16,533; 1980: 17,015; 1990: 16,799, according to New Jersey Department of Labor and Workforce Development, Office of Research and Information, "Table 6: New Jersey Resident Population by Municipality, 1930–1990", State of New Jersey, 1990, http://lwd.dol.state.nj.us/labor/lpa/census/1990/poptrd6.htm, and U.S. Census Bureau, Census 2000 and 2010: 16,930 and 2010: 16,116.
51. Ocean Grove Chamber of Commerce, "Ocean Grove: A Timeless Treasure at the Jersey Shore," 2014, http://www.oceangrovenj.com/.

52. Ocean Grove Chamber of Commerce, "About Ocean Grove," 2014, http://www.oceangrovenj.com/about.html.
53. City of Asbury Park, "City Guide" (2013), http://cityofasburypark.com/wp-content/uploads/2013/08/asbury_park_city_guide_20131.pdf, 1.
54. Gary J. Gates and Abigail M. Cooke, "New Jersey Census Snapshot: 2010," Williams Institute, UCLA School of Law, http://williamsinstitute.law.ucla.edu/wp-content/uploads/Census2010Snapshot_New-Jersey_v2.pdf.
55. Gary J. Gates and Abigail M. Cooke, "United States Census Snapshot: 2010," Williams Institute, UCLA School of Law, http://williamsinstitute.law.ucla.edu/wp-content/uploads/Census2010Snapshot-US-v2.pdf?r=1.
56. Human Rights Campaign Foundation, *Municipal Equality Index: A Nationwide Evaluation of Municipal Law* (Washington, DC: Human Rights Campaign Foundation, 2013), http://www.hrc.org/files/assets/resources/MEI_2013_report.pdf.
57. Eugene Paik, "Jenkinson's Owners to Repair Boardwalk So Bars Can Stay Open Past Midnight," *Star-Ledger*, December 20, 2012, http://www.nj.com/news/index.ssf/2012/12/jenkinsons_owners_to_repair_bo.html.
58. MaryAnn Spoto, "Ocean Grove Boardwalk Called a 'Miracle' Project after FEMA Funding Was Denied Twice," *Star-Ledger*, July 3, 2014, http://www.nj.com/monmouth/index.ssf/2014/07/ocean_grove_boardwalk_a_miracle_project_after_twice_denied_fema_funding.html.
59. MaryAnn Spoto, "State Sides with Lesbian Couple in Fight against Ocean Grove Association," *Star-Ledger*, December 30, 2008, http://www.nj.com/news/index.ssf/2008/12/judge_rules_monmouth_church_gr.html.
60. Christopher Robbins, "Christie Celebrates Asbury Park's Success at Boardwalk Ribbon Cutting," http://www.nj.com/monmouth/index.ssf/2014/05/christie_celebrates_asbury_parks_successes_at_boardwalk_ribbon-cutting.html; also see table 3.6.
61. Erin O'Neill, "Nearly $14M in Federal Sandy Aid Awarded to Asbury Park, Bradley Beach, and Sea Isle City," *Star-Ledger*, October 23, 2013, http://www.nj.com/news/index.ssf/2013/10/sandy_aid_sea_isle_asbury_park_bradley_beach.html.
62. Erin O'Neill, "'Awfully Empty': Seaside Heights Boardwalk Merchants Hold Out Hope for Better End to Summer," *Star-Ledger*, August 1, 2013, http://www.nj.com/ocean/index.ssf/2013/08/seaside_heights_boardwalk_business_sandy.html.
63. Borough of Seaside Heights, "2013 Municipal Data Sheet," 2013, http://www.seaside-heightsnj.org/wp-content/uploads/2010/07/MX-M453N_20130805_114731.pdf.
64. Associated Press, "Seaside Heights Municipal Court Revenues Dip $230K Due in Part to Sandy," http://www.nj.com/ocean/index.ssf/2013/02/seaside_heights_municipal_court_revenues_dip_240k_in_part_to_sandy.html.
65. Ed Beeson and Erin O'Neill, "Higher Flood Insurance Rates in Store for Thousands of New Jerseyans," *Star-Ledger*, March 29, 2013, http://www.nj.com/business/index.ssf/2013/03/higher_flood_insurance_rates_i.html.
66. U.S. Federal Emergency Management Agency (FEMA), "Rebuilding in a VE Zone," http://www.fema.gov/media-library/assets/documents/31594.
67. Stephen Sterling, "Jersey Shore Revolution Begins, as FEMA Releases New Flood Maps," *Star-Ledger*, December 16, 2012, http://www.nj.com/news/index.ssf/2012/12/jersey_shore_revolution_begins.html.
68. U.S. Federal Emergency Management Agency, "Flood Zones," accessed January 28, 2015, https://msc.fema.gov/webapp/wcs/stores/servlet/info?storeId=10001&catalogId=10001&langId=-1&content=floodZones&title=FEMA%2520Flood%2520Zone%2520Designations.

69. U.S. Federal Emergency Management Agency, "FEMA Flood Hazard Resources Map," accessed January 28, 2015, http://fema.maps.arcgis.com/home/webmap/viewer.html?webmap=2f0a884bfb434d76af8c15c26541a545&extent=-76.1688,38.4032,-70.4888,41.7479.
70. LBI House Raising, "Frequently Asked Questions," 2013, http://lbihouseraising.com/faqs/.

CHAPTER 4 THE SUBURBAN SHORE

1. Only Atlantic County contains a four-year public university (Richard Stockton State College). Monmouth and Ocean Counties have large two-year county colleges (Brookdale Community College and Ocean County College). Kean University (a four-year public university) offers some four-year degrees at the OCC campus. Atlantic and Cape May share Atlantic Cape Community College, which has three main campuses. The only private universities in the four Shore counties are Monmouth University in Long Branch, which enrolls about 5,600 students, and Georgian Court University in Lakewood (Ocean County), which enrolls about 2,200 students.
2. In no known universe would a train actually reach the Jersey Shore from stations in the order suggested; Bayonne is east of Newark but west of New York, Elizabeth is south of Newark, east of Bayonne, and south and east of New York. The New York and Long Branch rail line had only a branch line to Bayonne when it served Monmouth County; the Central Jersey railroad had routes that went from Jersey City to Newark and on to Elizabeth, or from Jersey City to Bayonne to Elizabeth, but none that began in Bayonne and then went to Elizabeth and then Newark. The contemporary New Jersey Transit North Jersey Coast line starts in New York and then travels to Newark and Elizabeth but does not make a stop in Bayonne. Thus, the term "Benny," while conceivably rooted in the names of North Jersey and New York cities, is unlikely to exactly reflect historic train tickets, as many locals claim.
3. Brad R. Tuttle, *How Newark Became Newark: The Rise, Fall, and Rebirth of an American City* (New Brunswick, NJ: Rutgers University Press, 2011).
4. William Julius Wilson, *The Truly Disadvantaged: The Inner City, the Underclass, and Public Policy* (Chicago: University of Chicago Press, 1987).
5. Elijah Anderson, *Code of the Street: Decency, Violence, and the Moral Life of the Inner City* (New York: W. W. Norton, 2000).
6. See Erin S. McCord and Jerry H. Ratcliffe, "A Micro-Spatial Analysis of the Demographic and Criminogenic Environment of Drug Markets in Philadelphia," *Australian and New Zealand Journal of Criminology* 40, no. 1 (2007): 43–63; and Kyle T. Bernstein, Sandro Galea, Jennifer Ahern, Melissa Tracy, and David Vlahov, "The Built Environment and Alcohol Consumption in Urban Neighborhoods," *Drug and Alcohol Dependence* 91 (2007): 244–252.
7. Joel Garreau, *Edge Cities: Life on the New Frontier* (New York: Doubleday, 1991).
8. Robert E. Lang, *Edgeless Cities: Exploring the Elusive Metropolis* (Washington, DC: Brookings Institute, 2003).
9. Herbert Spencer is credited with the idea and name of sociology as the scientific study of human society, but his theoretical work has been widely discredited as politically and socially elitist and fundamentally unscientific. Durkheim is the first sociologist to test empirical hypotheses with the systematic collection of data, most famously in his study *Suicide*, which compared suicide rates across countries with different religious traditions to identify a social component to this most individualized act. See Emile Durkheim, *Suicide: A Study in Sociology* (1951, New York: The Free Press).

10. Kingsley Davis and Wilber E. Moore, "Some Principles of Stratification," *American Sociological Review* 10, no. 2 (1970 [1945]): 242–249.
11. Thorstein Veblen, *The Theory of the Leisure Class: An Economic study of Institutions* (1899; repr., New York: Modern Library, 1934).
12. Pierre Bourdieu, *Distinction: A Social Critique of the Judgment of Taste* (Cambridge, MA: Harvard University Press, 1984).
13. Bryant Simon, *Boardwalk of Dreams: Atlantic City and the Fate of Urban America* (New York: Oxford University Press, 2004).
14. Douglas S. Massey et al., *Climbing Mount Laurel: The Struggle for Affordable Housing and Social Mobility in an American Suburb* (Princeton, NJ: Princeton University Press, 2013).
15. The Mount Laurel decisions by the New Jersey Supreme Court have attempted to alleviate this inequality by obligating each municipality to a regionally determined "fair share" of affordable housing. Massey et al. (*Climbing Mount Laurel*) detail how the legal framework can create effective housing alternatives in suburbia, but affordable housing obligations remain a highly contentious issue in the state and have failed to produce an economically integrated residential pattern in New Jersey. Massey's own research demonstrates that residential segregation has actually increased since the Mount Laurel decisions.
16. John R. Logan and Harvey Molotch, *Urban Fortunes: The Political Economy of Place*, 20th Anniversary Edition (1987; repr. Berkeley: University of California Press, 2007).
17. David K. Ihrke and Carol S. Faber, "Geographical Mobility, 2005–2010," U.S. Census Bureau (Dec. 2012), http://www.census.gov/prod/2012pubs/p20–567.pdf, 2.
18. Ibid., 11
19. Robert D. Bullard, "Solid Waste Sites and the Houston Black Community," *Sociological Inquiry* 53 (1983): 273–288.
20. Paul Mohai and Bunyan Bryant, "Environmental Injustice: Weighing Race and Class as Factors in the Distribution of Environmental Hazards," *University of Colorado Law Review* 63, no. 4 (1992): 921–932. Paul Mohai and Bunyan Bryant, eds., *Race and the Incidence of Environmental Hazards: A Time for Discourse* (Boulder, CO: Westview Press, 1992).
21. William J. Clinton, "Executive Order 12898 February 11, 1994: Federal Actions to Address Environmental Justice in Minority and Low-Income Populations," *Federal Register* 59, no. 32 (February 16, 1994), http://www.archives.gov/federal-register/executive-orders/pdf/12898.pdf.
22. Lisa Sun-Hee Park and David Naguib Pellow, *The Slums of Aspen: Immigrants vs. the Environment in America's Eden* (New York: New York University Press, 2011), 4.
23. Borough of Mantoloking, "How to . . . Enjoy Mantoloking's Beach" (2011), http://www.mantoloking.org/wp-content/uploads/2011/09/beachinfonew.pdf.
24. Local officials report 521 residential structures. The difference between local and Census officials is that some structures contained more than one housing unit.
25. New Jersey Department of Community Affairs, "Community Development Block Grant Disaster Recovery Action Plan. For CDBG-DR Disaster Recovery Funds, Disaster Relief Appropriations Act of 2013 (Public Law 113–2, January 29, 2013)," Trenton: State of New Jersey, 2013, http://www.state.nj.us/dca/divisions/sandyrecovery/pdf/CDBG-DisasterRecoveryActionPlan.%20non.substantial.amendments.%2011.14.13.pdf, 2–9.
26. The U.S. Census Bureau refers to Ortley Beach as "Dover Beaches South CDP," reflecting the former name of Toms River (Dover Township).
27. Joanna Peluso, "Framing Recovery: Understanding Inequality and Privilege during Hurricane Sandy Recovery in New Jersey" (senior sociology thesis, College of New Jersey, 2014).

CHAPTER 5 GOVERNMENT, BUREAUCRACY, AND TECHNICAL FIXES

1. Charles E. Fritz, "Disaster," in *Contemporary Social Problems: An Introduction to the Sociology of Deviant Behavior and Social Disorganization*, ed. Robert K. Merton and Robert A. Nisbet (New York: Harcourt, Brace, and World, 1961), 690–691.
2. Ibid., 690.
3. Known as the "State Mandate, State Pay" amendment, it can be found in article VIII, section II, paragraph 5 of the New Jersey State Constitution, http://www.njleg.state.nj.us/lawsconstitution/constitution.asp.
4. Douglas S. Massey et al., *Climbing Mount Laurel: The Struggle for Affordable Housing and Social Mobility in an American Suburb* (Princeton, NJ: Princeton University Press, 2013).
5. This also set off a sort of public theater in which the administration claimed to have lost the data on which the obligations were calculated, prompting one fair housing advocacy group to offer a reward for the missing documents. See Matt Friedman, "Affordable Housing Group Offers $1000 Reward for Document 'Lost' by Christie Administration," *Star-Ledger*, July 24, 2014, http://www.nj.com/politics/index.ssf/2014/07/affordable_housing_group_offers_1000_reward_for_document_lost_by_christie_administration.html#incart_river.
6. The smallest municipality in 2010 was Tavistock in Camden County, with just five residents according to the 2010 Census; U.S. Census Bureau, "American FactFinder," U.S. Census Bureau, 2010, http://factfinder2.census.gov.
7. Township of Toms River, "Toms River, New Jersey Departments," Township of Toms River, accessed February 2, 2015, http://tomsrivertownship.com/index.php/departments.html.
8. Middletown is not considered an oceanfront municipality in this analysis, as its ocean beach on Sandy Hook is part of the Gateway National Recreation Area, which is administered by the federal government.
9. CDPs do not always reflect local nomenclature. For example, the 2010 census identifies "Dover Beaches South CDP," reflecting neither the locally preferred "Ortley Beach" nor the 2006 official municipal name change from Dover Township to Toms River Township. Dover Beaches North CDP contains several locally recognized communities, such as Ocean Beach, Chadwick Island, and Normandy Beach.
10. Ocean Grove Homeowners Association, "Home," Ocean Grove Homeowners Association, accessed February 2, 2015, http://oghoa.org/.
11. Normandy Beach Improvement Association, "Welcome to the Normandy Beach Improvement Association, Inc. Website," Normandy Beach Improvement Association, accessed February 2, 2015, http://nbianj.org/.
12. Ortley Beach Voters and Tax Payers Association, "Home," Ortley Beach Voters and Tax Payers Association, 2014, http://www.ortleybeach.org/.
13. As an example, Ocean Beach I publishes its deed restrictions on-line: Ocean Beach Unit 1, "Club Documents," Ocean Beach Unit 1, 2015, http://www.obsc1.com/pages/clubDocuments.html.
14. Both figures from U.S. Census Bureau, 2010 Census.
15. Monmouth University/*Asbury Park Press* Poll, "Sandy's Impact on New Jersey," Monmouth University Polling Institute, December 10, 2012, http://www.monmouth.edu/assets/0/32212254770/32212254991/32212254992/32212254994/32212254995/30064771087/4de10cc0918e4afd8a6f6b8b0ff91aa6.pdf, 7.
16. Ibid.
17. FEMA calls these areas "Special Flood Hazard Areas," or SFHAs.

18. Stephen Stirling, "Jersey Shore Revolution Begins, as FEMA Releases New Flood Maps," *Star-Ledger*, December 16, 2012, http://www.nj.com/news/index.ssf/2012/12/jersey_shore_revolution_begins.html.
19. Peggy McGlone, "Sandy Victims Furious as FEMA Troubles Begin to Build," *Star-Ledger*, November 5, 2012, http://www.nj.com/news/index.ssf/2012/11/sandy_victims_furious_as_fema.html.
20. Eugene Paik, "4 N.J. Counties See Drastic Changes in Revised Flood Maps," *Star-Ledger*, June 18, 2013, http://www.nj.com/news/index.ssf/2013/06/after_new_flood_maps_scale_back_riskiest_areas_homeowners_in_4_nj_counties_make_decisions_on_elevati.html.
21. U.S. Federal Emergency Management Agency, "Floodsmart.gov: Know Your Risk, F-671 (10–12)" (Jessup, MD: U.S. Department of Homeland Security, 2012).
22. U.S. Census Bureau, 2010 Census
23. Ronda Kaysen, "Back to the Jersey Shore," *New York Times*, April 4, 2014, http://www.nytimes.com/2014/04/06/realestate/back-to-the-jersey-shore.html?_r=0.
24. Rutgers-Eagleton Poll, "Christie Gains as Smart, Effective Leader in Rutgers-Eagleton Poll," Eagleton Institute of Politics, November 29, 2012, http://eagletonpoll.rutgers.edu/polls/release_11–29–12.pdf.
25. Matt Friedman, "Fiery Christie Lashes Out at Boehner, House Republicans for Delaying Vote on Sandy Relief," *Star-Ledger*, January 2, 2013, http://www.nj.com/politics/index.ssf/2013/01/christie_sandy_relief_bill_pre.html.
26. See for example, the campaign's YouTube channel that features first-responders and victims of Sandy advocating for the governor, as well as footage of the governor touring communities devastated by the storm: https://www.youtube.com/user/ChristieforNJ.
27. MaryAnn Spoto, "State Adopts Flood Elevation Guidelines So Sandy Victims Can Rebuild Homes," *Star-Ledger*, January 24, 2013, http://www.nj.com/news/index.ssf/2013/01/post_299.html.
28. Michael Barbaro and Kate Zernike, "Mayor of Hoboken Says Hurricane Relief Was Threatened," *New York Times*, January 18, 2014, http://www.nytimes.com/2014/01/19/nyregion/florida-trip-is-no-reprieve-for-new-jersey-governor.html?_r=0.
29. Ted Sherman, "Sandy Housing Aid Went to Projects Far from Storm," *Star-Ledger*, February 16, 2014, http://www.nj.com/news/index.ssf/2014/02/sandy_housing_aid_went_to_projects_far_from_the_storm.html.
30. Monmouth University/*Asbury Park Press* Poll, "New Jersey Going Sour on Sandy Recovery," Monmouth University Polling Institute, April 16, 2014, http://www.monmouth.edu/assets/0/32212254770/32212254991/32212254992/32212254994/32212254995/30064771087/dc5583cc-eb3d-4dc7-bd54-e672e90cdf95.pdf, 4.
31. Ibid.
32. Monmouth University/*Asbury Park Press* Poll, "NJ Sandy Panel: Aid Applicants Yet to Feel 'reNEWed,'" Monmouth University Polling Institute, February 17, 2014, https://www.monmouth.edu/assets/0/32212254770/32212254991/32212254992/32212254994/32212254995/40802189893/FFE1103E4b3448088AA7e05a977ef0c7.pdf, 2.
33. See http://eagletonpoll.rutgers.edu/new-wp/wp-content/uploads/2014/04/release_04–23–14.pdf, 8.
34. Rutgers-Eagleton Poll, "Most New Jerseyans Still Think State Is Not Back to Normal Post-Sandy," Eagleton Institute of Politics, April 23, 2014, http://www.state.nj.us/gorr/plan/index.htmlhttps://docs.google.com/viewer?url=http://eagletonpoll.rutgers.edu/new-wp/wp-content/uploads/2014/04/release_04–23–14.pdf&chrome=true.

35. Hurricane Sandy New Jersey Relief Fund, "About Us," Hurricane Sandy New Jersey Relief Fund, accessed February 3, 2015, http://sandynjrelieffund.org/go.cfm?do=Page.View&pid=3.
36. New Jersey Office of Emergency Management, "Ready New Jersey: Alerts and Updates," New Jersey Office of Emergency Management, accessed February 3, 2015, http://readynj.wordpress.com/tag/pressreleases/.
37. See for example, the October 2012 Monmouth University/*Asbury Park Press* poll, which found that 47 percent of New Jersey residents surveyed rated the performance of first responders in Sandy as "excellent" and another 32 percent as "good": Monmouth University/*Asbury Park Press* Poll, "Sandy's Impact on New Jersey," Monmouth University Polling Institute, December10, 2012, http://www.monmouth.edu/assets/0/32212254770/32212254991/32212254992/32212254994/32212254995/30064771087/4de1occ0918e4afd8a6f6b8b0ff91aa6.pdf, 7.
38. New Jersey Office of Emergency Management, "Emergency Management in New Jersey," New Jersey Office of Emergency Management, accessed February 3, 2015, http://www.state.nj.us/njoem/press_emhistory.html.
39. Ibid.
40. New Jersey State Police, "Emergency Management Section," State of New Jersey, accessed February 3, 2015, http://www.njsp.org/divorg/homelandsec/ems.html.
41. Chris Christie, *The Governor's FY 2014 Budget* (Trenton: State of New Jersey Office of Management and Budget, 2014), http://www.state.nj.us/treasury/omb/publications/14budget/pdf/FY14BudgetBook.pdf,322.
42. For a firsthand account of being a research scientist for the New Jersey Department of Environmental Protection, see Thomas Belton, *Protecting New Jersey's Environment: From Cancer Alley to the New Garden State* (New Brunswick, NJ: Rutgers University Press, 2011).
43. N.J. Stat. § 13:9A-1 et seq.
44. N.J. Stat. § 13:19–1 et seq.
45. N.J. Stat. § 13:9A-4.
46. N.J.A.C. 7:7E-1.1(c).
47. Christie, *The Governor's FY 2014 Budget*, 183. Notably, only three state organizations (DEP, Department of Transportation, and Law Enforcement) mentioned Sandy or Sandy recovery in their official budget narratives. All three saw their budgets reduced from 2013 to 2014, with the Department of Transportation registering the greatest decrease ($75 million to $44 million), as a result of the Sandy-related construction that ballooned in the DOT's 2013 budget.
48. New Jersey Department of Community Affairs, "About DCA," State of New Jersey Department of Community Affairs, 2015, http://www.state.nj.us/dca/about/index.html.
49. New Jersey Department of Community Affairs, "Community Development Block Grant Disaster Recovery Action Plan. For CDBG-DR Disaster Recovery Funds, Disaster Relief Appropriations Act of 2013 (Public Law 113–2, January 29, 2013)," 2013, http://www.state.nj.us/dca/divisions/sandyrecovery/pdf/CDBG-DisasterRecoveryActionPlan.%20non.substantial.amendments.%2011.14.13.pdf.
50. State of New Jersey Office of the State Comptroller, "NJ Sandy Transparency," State of New Jersey, 2015, http://nj.gov/comptroller/sandytransparency/.
51. NJ DCA (2013), Table 2–8.
52. William R. Freudenberg, "Risk and Recreancy: Weber, the Division of Labor, and the Rationality of Risk Perceptions," *Social Forces* 71, no. 4 (1993): 909–932, quotation on 916–917.

53. Valerie Gunter and Steve Kroll-Smith, *Volatile Places: A Sociology of Communities and Environmental Controversies* (Thousand Oaks, CA: Pine Forge Press, 2007), 72–73.
54. Ibid., 96.
55. Jennifer L. Irish, Patrick J. Lynett, Robert Weiss, Stephanie M. Smallegan, and Wei Cheng, "Buried Relic Seawall Mitigates Hurricane Sandy's Impact," *Coastal Engineering* 80 (2013): 79–82.
56. MaryAnn Spoto, "Two N.J. Shore Towns Divided over New Sea Wall," *Star-Ledger*, March 10, 2013, http://www.nj.com/ocean/index.ssf/2013/03/mantoloking_bay_head_sandy.html.
57. U.S. Army Corps of Engineers, "New Jersey Shore Protection, Manasquan Inlet to Barnegat Inlet, NJ (Fact Sheet)," U.S. Army Corps of Engineers, 2014, http://www.nap.usace.army.mil/Missions/Factsheets/FactSheetArticleView/tabid/4694/Article/490786/new-jersey-shore-protection-manasquan-inlet-to-barnegat-inlet-nj.aspx.
58. Ryan Hutchins, "Gov. Christie Says N.J. Will Build Dune System in Sandy's Wake," *Star-Ledger*, April 30, 2013, http://www.nj.com/politics/index.ssf/2013/04/gov_christie_says_nj_will_buil.html.
59. U.S. Army Corps of Engineers, "Sea Bright to Manasquan, NJ (Fact Sheet)," U.S. Army Corps of Engineers, 2015, http://www.nan.usace.army.mil/Media/FactSheets/FactSheetArticleView/tabid/11241/Article/487661/sea-bright-to-manasquan-nj-beach.aspx.
60. U.S. Army Corps of Engineers, "New Jersey Shore Protection, Manasquan Inlet to Barnegat Inlet, NJ (Fact Sheet)," and U.S. Army Corps of Engineers, "New Jersey Shore Protection, Barnegat Inlet to Little Egg Inlet (Long Beach Island, NJ) (Fact Sheet)," U.S. Army Corps of Engineers, 2014, http://www.nap.usace.army.mil/Missions/Factsheets/FactSheetArticleView/tabid/4694/Article/490783/new-jersey-shore-protection-barnegat-inlet-to-little-egg-inlet-long-beach-islan.aspx.
61. U.S. Army Corps of Engineers, "New Jersey Shore Protection, Barnegat Inlet to Little Egg Inlet, (Long Beach Island, NJ) (Fact Sheet)."
62. State of New Jersey Department of Environmental Protection, "Blue Acres Success Stories," State of New Jersey, 2014, http://www.nj.gov/dep/greenacres/blue_success.html.
63. Kate Millsaps, "One Year after Sandy Changes Still Needed," press release, Sierra Club New Jersey Chapter, October 21, 2013, http://newjersey.sierraclub.org/PressReleases/0520.asp.
64. Lixion Avila and John Cangialosi, "Tropical Cyclone Report: Hurricane Irene (AL092011) 21–28 August 2011," National Oceanic and Atmospheric Agency National Hurricane Center, 2012, http://www.nhc.noaa.gov/data/tcr/AL092011_Irene.pdf, 5-6.
65. New Jersey Department of Community Affairs, "New Jersey Community Development Block Grant Disaster Recovery Action Plan State Fiscal Year 2013," State of New Jersey, http://www.nj.gov/dca/divisions/dhcr/rfp/pdf/cdbg_dr_plan_rfp_gl.pdf, 8.
66. "N.J. Braces for Hurricane Irene's Wrath as Officials Warn of High Winds, Heavy Flooding, and Possible Power Outages," *Star-Ledger*, August 26, 2011, http://www.nj.com/news/index.ssf/2011/08/nj_braces_for_hurricane_irenes_1.html.
67. National Oceanic and Atmospheric Administration, "Service Assessment: Hurricane Irene, August 21–30, 2011," U.S. Department of Commerce, http://www.nws.noaa.gov/om/assessments/pdfs/Irene2012.pdf, xi, 65.
68. Christopher Baxter, "Gov. Christie Blasts A.C. Mayor for Keeping Residents in Path of Hurricane Sandy," *Star-Ledger*, October 29, 2012, http://www.nj.com/news/index.ssf/2012/10/gov_christie_blasts_ac_mayor_f.html.

CHAPTER 6 RESTORING SECURITY AT THE SHORE

1. Anthony Giddens, *The Consequences of Modernity* (Stanford, CA: Stanford University Press, 1990). Also see: Anthony Giddens, *Modernity and Self-Identity: Self and Society in the Late Modern Age* (Stanford, CA: Stanford University Press, 1991).
2. Cited in New Jersey Department of Community Affairs, "Community Development Block Grant Disaster Recovery Action Plan. For CDBG-DR Disaster Recovery Funds, Disaster Relief Appropriations Act of 2013 (Public Law 113–2, January 29, 2013)," http://www.state.nj.us/dca/divisions/sandyrecovery/pdf/CDBG-DisasterRecoveryActionPlan.%20non.substantial.amendments.%2011.14.13.pdf. 2–3.
3. Ibid., 2–11.
4. Dan Goldberg, "Seaside Heights Residents Allowed to Return to Their Homes Friday," *Star-Ledger*, November 8, 2012, http://www.nj.com/news/index.ssf/2012/11/seaside_heights_residents_allo.html.
5. Monmouth University/*Asbury Park Press* Poll, "NJ Sandy Panel: Mental Health Issues Persist," Monmouth University Polling Institute, March 20, 2014, http://www.monmouth.edu/assets/0/32212254770/32212254991/32212254992/32212254994/32212254995/30064771087/414676ee-2ecb-4eb5-a3ce-e9cd0b18bf0c.pdf.
6. Atlantic City Electric, "Atlantic City Electric Restores Power to All Customers Affected by Hurricane Sandy Who Can Safely Accept Electric," Atlantic City Electric press release, November 6, 2012, http://www.atlanticcityelectric.com/library/templates/Interior.aspx?Pageid=1105&id=15258.
7. Kayla Webley, "Hurricane Sandy by the Numbers: A Superstorm's Statistics, One Month Later," *Time*, November 26, 2012, http://nation.time.com/2012/11/26/hurricane-sandy-one-month-later/.
8. PSE&G, "PSE&G Storm Update—Tuesday, October 30, 2012 at 4 A.M.," PSE&G press releases, Oct. 30, 2012, https://www.pseg.com/info/media/newsreleases/2012/2012–10–30a.jsp.
9. PSE&G, "PSE&G Service Restoration Update—Friday, November 9 at 11:45 A.M.," PSE&G press release, November 9, 2012, https://www.pseg.com/info/media/newsreleases/2012/2012–11–09a.jsp.
10. Ibid.
11. Julia Terruso, Christopher Baxter, and Eliot Caroom, "Sandy Recover Becomes National Mission as Countless Workers come to N.J.'s Aid," *Star-Ledger*, November 11, 2012, http://www.nj.com/news/index.ssf/2012/11/sandy_relief_becomes_national.html.
12. *Star-Ledger*, *When Sandy Hit: The Storm that Forever Changed New Jersey* (Battle Ground, WA: Pediment Publishing, 2013).
13. Dave Fried, "JCP&L, FirstEnergy Fall Far Short in N.J.: Opinion," *Star-Ledger*, November 20, 2012, http://blog.nj.com/njv_guest_blog/2012/11/jcpl_firstenergy_fall_far_shor.html.
14. Amy Brittain and Richard Khavkine, "Gas Panic: Sandy Sparks Long Lines, Short Tempers as Residents Scramble to Fill Up," *Star-Ledger*, November 1, 2012, http://www.nj.com/news/index.ssf/2012/11/gas_panic_sandy_sparks_long_li.html.
15. New Jersey Office of the Governor, "Christie Administration Investigates Complaints of Alleged Storm-Related Price Gouging," press release, State of New Jersey, October 31, 2012, http://www.nj.gov/oag/newsreleases12/pr20121031a.pdf; New Jersey Division of Consumer Affairs, "Christie Administration Subpoenas 65 Businesses in Investigations into Post-Hurricane Price Gouging," State of New Jersey Department of Law & Public Safety, November 2, 2012, http://www.njconsumeraffairs.gov/press/11022012.htm; New

Jersey Division of Consumer Affairs, "Christie Administration Files New Price Gouging Lawsuits against Ten More New Jersey Businesses, Including Seven Hotels Collectively Accused of More Than 1,000 Instances of Price Gouging, Following Hurricane Sandy," State of New Jersey Department of Law & Public Safety, November 28, 2012, http://www.njconsumeraffairs.gov/press/11282012.htm.

16. New Jersey Division of Consumer Affairs, "Alleged Price Gougers Will Pay Over $430,000 to Settle State Lawsuits Filed Following Superstorm Sandy; Total Recoveries in Sandy Gouging Cases Top $800,000," State of New Jersey Department of Law & Public Safety, October 30, 2013, http://www.njconsumeraffairs.gov/press/10302013b.pdf.
17. New Jersey Office of the Governor, "Christie Administration Warns Consumers to Beware of Home Repair Scams, Charity Scams While Recovering from Storm Damage," press release, November 1, 2012, http://www.state.nj.us/governor/news/news/552012/approved/20121101g.html.
18. New Jersey Office of the Governor, "New Jersey Division of Consumer Affairs Issues Warning to Organizations Soliciting Charitable Donations for Hurricane Sandy Victims," press release, November 1, 2012, http://www.njconsumeraffairs.gov/press/02112013.htm.
19. The JSHN has over 200,000 followers and is now well networked into traditional media. JSHN's founder, amiable city planner Justin Auciello, has become the voice of the Jersey Shore, sought after by local and national media. For example, when the fire decimated the Funtown Pier in September 2013, Auciello was interviewed by NBC, the *Star-Ledger*, and *USA Today*, and provided reporting for several New York and Philadelphia television stations, among others. JSHN can be found at: https://www.facebook.com/JerseyShoreHurricaneNews.
20. *Star-Ledger, When Sandy Hit.*
21. Lee Clarke, "Like a Hurricane—but Worse," *Contexts* 12, no. 1 (2013): 7.
22. Rutgers-Eagleton Poll, "Most New Jerseyans Still Think State Is Not Back to Normal Post-Sandy," Eagleton Institute of Politics, April 23, 2014, https://docs.google.com/viewer?url=http://eagletonpoll.rutgers.edu/new-wp/wp-content/uploads/2014/04/release_04-23-14.pdf&chrome=true.
23. Kai T. Erikson, *Everything in its Path: Destruction of Community in the Buffalo Creek Flood* (New York: Simon & Schuster Paperbacks, 1976).
24. Duane A. Gill and J. Steven Picou, "Technological Disaster and Chronic Community Stress," *Society & Natural Resources*, 11, no. 8 (1998).
25. Giddens, *The Consequences of Modernity*, 90.
26. Ibid., 127.
27. Lee Clarke, Lee, *Worst Cases: Terror and Catastrophe in the Popular Imagination* (Chicago: University of Chicago Press, 2003).
28. Giddens, *Consequences of Modernity*, 134–137.
29. Ulrich Beck, *Risk Society: Towards a New Modernity* (Thousand Oaks, CA: Sage Publications, 1992).
30. Beck does not claim that material scarcity has disappeared in a risk society, and he concedes that one's exposure to risk is affected by one's material, or class, position. However, a focus on material scarcity tends to distract people from the accumulation of risks, which doesn't alleviate risks and can actually make people more vulnerable to them. Moreover, material scarcity tends to reduce political power to define and assert risks.
31. The probability of winning is exactly the same with each spin, no matter how many spins a person takes.

32. Kari Marie Norgaard, *Living in Denial: Climate Change, Emotions, and Everyday Life* (Cambridge, MA: MIT Press, 2011).
33. "Nonrational" here is used in a Weberian sense and should not be confused with "irrational." Weberian rationality is predicated on value-neutral, utilitarian (often economic) calculations. Nonrational behavior relies on social and emotional logics. For example, raising children in modern society is not strictly rational for individual parents, who must forgo considerable personal expense with limited material return. Still, people have children because they have social and emotional reasons for doing so.
34. Technically, the Sandy Recovery Improvement Act, which is Division B of the Disaster Relief Appropriations Act of 2013.
35. New Jersey Department of State, Division of Travel and Tourism, *New Jersey* 2014 *Travel Guide* (Trenton: State of New Jersey, 2014), http://www.visitnj.org/sites/default/master/files/New-Jersey-2014-Vistors-Guide-web-lo.pdf, 5.
36. Ibid., 23.
37. The Sawmill, "The Sawmill," accessed February 4, 2014, http://www.sawmillcafe.com/.
38. William Sokolic, "Jersey Shore Tourism Shines in Sunny Summer of 2014," *Courier Post* (Cherry Hill, NJ), September 1, 2014, http://www.courierpostonline.com/story/news/local/south-jersey/2014/08/30/blue-skies-bright-eyes/14877865/.
39. Patrick M. Harrington, "Devaluation and Developing Law: An Analysis of Coastal Property Rights and Eminent Domain in New Jersey in the Wake of Super Storm Sandy and the Recent New Jersey Supreme Court Decision in Harvey Cedars v. Karan (August 6, 2013)," available at SSRN: http://ssrn.com/abstract=2362543 or http://dx.doi.org/10.2139/ssrn.2362543.
40. MaryAnn Spoto, "Harvey Cedars Couple Receives $1 Settlement for Dune Blocking Ocean View," *Star-Ledger*, September 25, 2013, http://www.nj.com/ocean/index.ssf/2013/09/harvey_cedars_sand_dune_dispute_settled.html.
41. See, for example, Rutgers Climate Institute, *State of the Climate: New Jersey* 2014 (New Brunswick: Rutgers Climate Institute, 2014), http://climatechange.rutgers.edu/custom/climatereport-final-2013/.
42. As of October 2014, this was still a possibility. See Lisa Rose, "Seaside Park's FunTown Amusement Pier, Destroyed in Fire, May Be Rebuilt by Next Summer," *Star-Ledger*, October 28, 2014, http://www.nj.com/ocean/index.ssf/2014/10/funtown_pier_may_be_back_for_next_summer_with_new_rides_owner_says.html.

SELECTED BIBLIOGRAPHY

Beck, Ulrich. *Risk Society: Towards a New Modernity.* Thousand Oaks, CA: Sage Publications, 1992.

Belton, Thomas. *Protecting New Jersey's Environment: From Cancer Alley to the New Garden State.* New Brunswick, NJ: Rutgers University Press, 2011.

Bourdieu, Pierre. *Distinction: A Social Critique of the Judgment of Taste.* Cambridge, MA: Harvard University Press, 1984.

Bullard, Robert D. *Dumping in Dixie: Race, Class, and Environmental Quality.* Boulder, CO: Westview Press, 2000 [1990].

———. "Solid Waste Sites and the Houston Black Community." *Sociological Inquiry* 53 (1983): 273–288.

Catton, William R. *Overshoot: The Ecological Basis of Revolutionary Change.* Urbana: University of Illinois Press, 1980.

Catton, William R., and Riley E. Dunlap. "Environmental Sociology: A New Paradigm." *American Sociologist* 13 (1978): 41–49.

Clarke, Lee. *Worst Cases: Terror and Catastrophe in the Popular Imagination.* Chicago: University of Chicago Press, 2003.

Davis, Kingsley, and Wilber E. Moore. "Some Principles of Stratification." *American Sociological Review* 10, no. 2 (1970 [1945]): 242–249.

Dunlap, Riley, and Angela Mertig. *American Environmentalism: The U.S. Environmental Movement, 1970–1990.* New York: Routledge, 1992.

Erikson, Kai T. *Everything in Its Path: Destruction of Community in the Buffalo Creek Flood.* New York: Simon & Schuster Paperbacks, 1976.

Florida, Richard. *The Rise of the Creative Class and How It's Transforming Work, Leisure, and Everyday Life.* New York: Basic Books, 2002.

Foster, John Bellamy. *Ecology Against Capitalism.* New York: Monthly Review Press, 2002.

Freudenberg, William R. "Risk and Recreancy: Weber, the Division of Labor, and the Rationality of Risk Perceptions." *Social Forces* 71, no. 4 (1993): 909–932.

Freudenberg, William R., and Robert Gramling. *Blowout in the Gulf: The BP Oil Spill Disaster and the Future of Energy in America.* Cambridge, MA: MIT Press, 2010.

Garreau, Joel. *Edge Cities: Life on the New Frontier.* New York: Doubleday, 1991.

Giddens, Anthony. *The Consequences of Modernity.* Stanford, CA: Stanford University Press, 1990.

———. *Modernity and Self-Identity: Self and Society in the Late Modern Age.* Stanford, CA: Stanford University Press, 1991.

Gunter, Valerie, and Steve Kroll-Smith. *Volatile Places: A Sociology of Communities and Environmental Controversies.* Thousand Oaks, CA: Pine Forge Press, 2007.

Hannigan, John. *Environmental Sociology.* 2nd ed. New York: Routledge, 2006.

Lang, Robert E. *Edgeless Cities: Exploring the Elusive Metropolis.* Washington, DC: Brookings Institute, 2003.

Logan, John R., and Harvey Molotch. *Urban Fortunes: The Political Economy of Place.* 1987. 20th anniversary edition. Berkeley: University of California Press, 2007.

Massey, Douglas S., Len Albright, Rebecca Casciano, Elizabeth Derickson, and David N. Kinsey. *Climbing Mount Laurel: The Struggle for Affordable Housing and Social Mobility in an American Suburb.* Princeton, NJ: Princeton University Press, 2013.

Mohai, Paul, and Bunyan Bryant, eds. *Race and the Incidence of Environmental Hazards: A Time for Discourse.* Boulder, CO: Westview Press, 1992.

Molotch, Harvey. "Oil in Santa Barbara and Power in America." *Sociological Inquiry* 40 (1970): 131–144.

Norgaard, Kari Marie. *Living in Denial: Climate Change, Emotions, and Everyday Life.* Cambridge, MA: MIT Press, 2011.

Park, Lisa Sun-Hee, and David Naguib Pellow. *The Slums of Aspen: Immigrants vs. the Environment in America's Eden.* New York: New York University Press, 2011.

Rudel, Thomas K., with Bruce Horowitz. *Tropical Deforestation: Small Farmers and Land Clearing in the Ecuadorian Amazon.* New York: Columbia University Press, 1993.

Salter, Edwin. *A History of Monmouth and Ocean Counties: Embracing a Geneological Record of Earliest Settlers in Monmouth and Ocean Counties and Their Descendants. The Indians: Their Language, Manners, and Customs. Important Historical Events: The Revolutionary War, Battle of Monmouth, the War of the Rebellion: Names of Officers and Men of Monmouth and Ocean Counties Engaged in It, etc., etc.* Bayonne, NJ: E. Gardner & Son, Publishers, 1890.

Schaiberg, Allan. *The Environment: From Surplus to Scarcity.* New York: Oxford University Press, 1980.

Simon, Bryant. *Boardwalk of Dreams: Atlantic City and the Fate of Urban America.* New York: Oxford University Press, 2004.

Stansfield, Charles A. *A Geography of New Jersey: The City in the Garden.* 2nd ed. New Brunswick, NJ: Rutgers University Press, 2004.

Star-Ledger. When Sandy Hit: The Storm That Forever Changed New Jersey. Battle Ground, WA: Pediment Publishing, 2013.

Tuttle, Brad R. *How Newark Became Newark: The Rise, Fall, and Rebirth of an American City.* New Brunswick, NJ: Rutgers University Press, 2011.

Veblen, Thorstein. *The Theory of the Leisure Class: An Economic Study of Institutions.* New York: Modern Library, 1934 [1899].

Weslager, C. A. *The Delaware Indians: A History.* New Brunswick, NJ: Rutgers University Press, 1990.

Wilson, William Julius. *The Truly Disadvantaged: The Inner City, the Underclass, and Public Policy.* Chicago: University of Chicago Press, 1987.

Wolff, Daniel J. *4th of July, Asbury Park: A History of the Promised Land.* New York: Bloomsbury Publishing, 2005.

Zukin, Sharon. *Naked City: The Death and Life of Authentic Urban Places.* New York: Oxford University Press, 2011.

INDEX

Note: Page numbers in italics indicate photos or maps; *t* indicates a table.

ABOUT THE AUTHOR

DIANE C. BATES is a professor of sociology and coordinator of environmental studies at The College of New Jersey. She is the author of numerous peer-reviewed articles and book chapters investigating the relationship between human society and the nonhuman environment.

CPSIA information can be obtained
at www.ICGtesting.com
Printed in the USA
LVOW13s0751060917
547623LV00018B/259/P

9 780813 57339